Expression Systems and Processes for rDNA Products

Expression Systems and Processes for rDNA Products

Randolph T. Hatch, EDITOR
Aaston, Inc.

Charles Goochee, EDITOR
Stanford University

Antonio Moreira, EDITOR
University of Maryland

Yair Alroy, EDITOR
Schering-Plough Corporation

Developed from symposia sponsored
by the Division of Biochemical Technology
of the American Chemical Society

American Chemical Society, Washington, DC 1991

SEPIAE
CHEM

Library of Congress Cataloging-in-Publication Data

Expression systems and processes for rDNA products /   Randolph T. Hatch, editor ... [et al.].

p.   cm.—(ACS symposium series; 477).

"Developed from a symposium sponsored by the Division of Biochemical Technology of the American Chemical Society."

Includes bibliographical references and index.

ISBN 0–8412–2172–3

1. Recombinant proteins—Synthesis. 2. Gene expression. 3. Genetic engineering.

I. Hatch, Randolph T., 1945–   . II. American Chemical Society. Division of Biochemical Technology. III. Series.

TP248.65.P76E9   1991
660'.65—dc20   91–36696
                                    CIP

The paper used in this publication meets the minimum requirements of American National Standard for Information Sciences—Permanence of Paper for Printed Library Materials, ANSI Z39.48–1984.   ∞

Foreword

THE ACS SYMPOSIUM SERIES was founded in 1974 to provide a medium for publishing symposia quickly in book form. The format of the Series parallels that of the continuing ADVANCES IN CHEMISTRY SERIES except that, in order to save time, the papers are not typeset, but are reproduced as they are submitted by the authors in camera-ready form. Papers are reviewed under the supervision of the editors with the assistance of the Advisory Board and are selected to maintain the integrity of the symposia. Both reviews and reports of research are acceptable, because symposia may embrace both types of presentation. However, verbatim reproductions of previously published papers are not accepted.

Contents

INDEXES

Preface

THE USE OF BACTERIAL HOSTS such as *Escherichia coli* for the production of recombinant proteins has many limitations. One set of problems relates to the intracellular accumulation of the product: The total achievable product titer is limited by the achievable cell density; the protein is frequently produced in a denatured state and therefore requires subsequent costly and inefficient renaturation; and the bacterial membrane has to be ruptured (or the whole cell has to be broken) to release the protein, complicating downstream purification procedures. An additional problem associated with intracellular accumulation of proteins is the requirement that the protein be compatible with the host cell. This requirement further restricts the range of proteins that can be produced by a given bacterial host. A problem associated with the use of procaryotic bacterial hosts, such as *E. coli*, is the inability to glycosylate proteins. Although many proteins do not require glycosylation for activity, proteins originating in higher life forms (such as animal, insect, and plant tissues) often require glycosylation. Finally, for the product to be acceptable for use in foods and beverages, the host bacteria should be generally recognized as safe (GRAS), which is not the case for *E. coli*, to ease regulatory approval of the production process.

For many reasons, including those just given, it has become important to expand the range of hosts available for production of products. The rapid development of the tools of genetic engineering for the manipulation of cellular DNA has provided the necessary capability to develop a number of successful applications in an expanding range of expression systems.

Expression systems, which originated in procaryotic bacteria with internally accumulated proteins encoded by bacterial DNA, have evolved to increasingly complex, higher life forms. It is now possible to produce proteins encoded by the DNA of foreign organisms in a wide variety of procaryotic and eucaryotic cells (including yeast, insect, plant, and animal cells). The technology of genetic engineering has also led to routine use of new expression systems for the synthesis and secretion of proteins in prokaryotic microorganisms with DNA originating from eukaryotic cells. This technology includes production systems capable of yielding high concentrations of secreted proteins (1–2 g/L) in GRAS hosts. It has also led to a number of production systems capable of glycosylating proteins.

Since in vitro manipulation of DNA through the use of polymerase chain reaction (PCR) was developed by Cetus, the utility of recombinant expression systems for the production of useful proteins has been significantly expanded. With PCR, it is now possible to modify DNA rapidly to alter the structure and properties of proteins to suit specific applications. The ease of production of test quantities of the new proteins then allows the determination of the structure–function relationship in the environment of the application and the subsequent development of proteins with properties optimally suited to the end use. This capability, coupled with the availability of a wide range of expression systems, is leading to an ever-increasing range of applications, including those in the pharmaceutical, food, and chemical industries.

To improve future production systems further, it will be necessary to determine the dynamics of the intracellular processes. The rate-limiting steps are still inadequately characterized to permit full optimization of the production and secretion of proteins. It is also necessary to understand the energetics of intracellular processes further to allow optimization of small molecule production.

The purpose of this volume is to report on recent developments in new expression system technologies as well as relevant process technology. The chapters discuss bacterial hosts (*E. coli*), yeast (*Saccharomyces cerevisiae*), and insect cells. The process technologies included are high-cell-density bacterial fermentations, biocatalysis, and process issues with recombinant microorganisms. Overall, the book represents the range of activities under way in the development and use of recombinant microorganisms and tissues for advanced processes and products.

Acknowledgments

We express our appreciation to the authors for their contributions to this book. We also thank the Division of Biochemical Technology for sponsoring the symposia represented by these papers. Finally, we are especially appreciative to the editorial staff of the ACS Books Department and to Robin Giroux.

RANDOLPH T. HATCH
Aaston, Inc.
12 Falmouth Road
Wellesley, MA 02181

ANTONIO MOREIRA
University of Maryland
TRC, Room 250
Baltimore, MD 21228

CHARLES GOOCHEE
Department of Chemical Engineering
Stanford University
Stanford, CA 94305–5025

YAIR ALROY
Schering-Plough Corporation
1011 Morris Avenue
Union, NJ 07083

September 2, 1991

Chapter 1

Secondary Concerns of Recombinant Microorganism Processing

Steven J. Coppella[1,2,4], Gregory F. Payne[1,3], and Neslihan DelaCruz[1,5]

[1]Chemical and Biochemical Engineering Program, University
of Maryland—Baltimore County, TRC Building, Baltimore, MD 21228
[2]Medical Biotechnology Center, Maryland Biotechnology Institute,
University of Maryland—Baltimore, Baltimore, MD 21201
[3]Agricultural Biotechnology Center, Maryland Biotechnology Institute,
University of Maryland, College Park, MD 20740

This paper addresses concerns of recombinant microorganism processing including genetic and nongenetic problems. The nongenetic problems discussed include physiological concerns (medium effects, nonoxidative product formation in an aerobic environment, oxygen supply limitations), and product instabilities. The impact of these effects on the host growth and product formation will be shown to be both significant and varied thereby necessitating individual study.

The study of biological phenomena follows a progression of questions which begin with the most critical then progresses to "secondary concerns". Certainly if one examines the research of recombinant microorganism processing, this pattern emerges. With the advent of recombinant DNA techniques for *E. coli* in the late 1970's, the first concern was with the techniques of DNA manipulation (e.g. digestion, ligation, annealing, etc.), followed by the design of the plasmid (origin of replication, selectable marker, etc.). After stable constructions became available, research could be focussed on the effects of the environment on the maintenance of the plasmid and expression of recombinant proteins. This same pattern developed for each new host as their genetics became understood (e.g. yeast, *Streptomyces*, *Bacillus*, etc.). We will attempt to discuss and characterize these secondary concerns for several host systems and try to generalize what can be learned.

For the purposes of discussion, we will be pragmatic and define the product formation to be the effect and categorize causes of loss in product formation into Genetic Problems and Nongenetic Problems. We will further categorize Nongenetic Problems in terms of Physiology (Medium Effects, Nonoxidative Product Formation

[4]Current address: Pure Carbon Company, 441 Hall Avenue, St. Marys, PA 15857
[5]Current address: Arctech Inc., 5390 Cherokee Avenue, Alexandria, VA 22312

0097–6156/91/0477–0001$06.00/0

under aerobic conditions, and Oxygen Transfer Limitations) and Product Degradation as illustrated in Figure 1. This figure should provide a road map for the following discussion and hopefully keep the issues discussed in perspective. We will not discuss the characterization of product synthesis rates (e.g. growth associated or nongrowth associated production kinetics) which would certainly require a Symposium Series of its own.

GENETIC PROBLEMS

PLASMID MAINTENANCE: CATEGORIZING PLASMID LOSS.

Losses in the number of "active" plasmids per cell results from several occurrences: (1) a decreased rate in plasmid replication relative to cell replication which would decrease the plasmid copy number; (2) insertion or deletion (structural stability) in a region of the plasmid required for replication or segregation, or protein expression (selective marker gene or product gene); (3) plasmid multimerization which could affect the plasmid copy number; and (4) inefficient partitioning of plasmids at cell division (segregation stability) leaving progeny cells with a decreased number of, or no, plasmids. The maintaining of a stable copy number through the control of copy number and plasmid partitioning will be referred to as "plasmid maintenance".

Variability in the plasmid copy number, segregation stability, and structural stability have been observed for many plasmids and hosts and these are influenced by a variety of environmental parameters. Segregation stability and copy number in *E. coli* can be strongly dependent on the plasmid construction (1,2,3,4), growth rate (5,6), and temperature (7). Nugent et al. (8) found the copy number and structural stability of plasmids in *E. coli* to vary significantly thus necessitating methods for the rapid and quantitative measurement of plasmid copy number and integrity. Structural stability of plasmids in *Bacillus* can also be highly dependent on plasmid construction, and plasmid content on temperature and specific growth rate (5). In yeast the plasmid copy number and plasmid segregation stability can also be strongly dependent on the host (9,10,11), plasmid construction, the presence of endogenous plasmids (12) and selection for gene products encoded on the plasmid (13).

For a given host the genetic problems will be a direct result of the plasmid construction and the environment. This construction can include the need for several regions involved in plasmid replication and partitioning. Selection is required for periodic determination of correct phenotype and is convenient for determining the fraction of recombinant cells. "Active selection", including antibiotic resistance, is undesirable because it adversely affects cell growth and protein synthesis rates, is costly, and yeast are normally resistant to most antibiotics. "Passive selection" includes selecting for the ability to synthesize an amino acid or nucleotide included in the plasmid construction that is deficient in the selected host. By depriving a medium of this requirement, recombinant cells are passively selected. Three different passive selections are commonly used in yeast: the ability to synthesize uridine, histidine, and leucine. Hollenberg (14) reported selection for leucine with the Leu2 allele yields the highest copy number with the 2 μm origin, and so leucine selection is often the selection method of choice. As will be seen later in the discussion of medium effects, selective pressure can have a strong effect on cell growth and protein production.

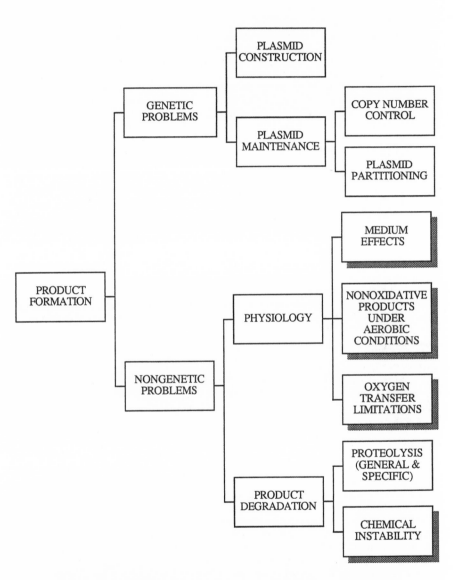

Figure 1. Relationships between the various factors which can affect product formation. The shaded regions comprise the focus of this manuscript.

DIFFICULTIES IN DEFINING THE GENETIC PROBLEMS.

One of the problems of genetic instability is in defining the way in which this instability is to be measured as it is the measurement that must be interpreted to draw conclusions on the plasmid instability. Four prominent methods of measuring plasmid stability are: (1) determining the % recombinant cells by exploiting a phenotype conferred by the plasmid; (2) measuring bulk copy number using cesium chloride/ethidium bromide gradient centrifugation of total DNA, gel electrophoresis, HPLC or hybridization to a specific DNA probe; (3) using Flow Cytometry to measure a protein produced from the plasmid; and (4) following protein production over time. The determination of % recombinant cells is usually done using the selective marker. Cell samples are diluted in a saline solution and plated onto a nonselective agar medium. The colonies are allowed to grow then replica plated onto a selective agar medium and the % recombinant cells calculated. Although a simple assay to perform, this method of determining plasmid stability neglects any structural instability in other product genes and therefore also requires the product protein assay(s) as a supporting measurement.

Bulk copy number determinations using HPLC or hybridization to a labeled probe do not account for structural instability of the plasmid which could strongly affect product formation. These two methods do not measure the distribution of plasmid copy number in the cells which is the motivation for Flow Cytometry. This latter technique uses a marker protein product which can be made to fluorescence and level of the fluoresce measured for individual cells. The fluorescence levels are correlated to protein concentration which in turn are correlated to plasmid copy number. Although the method gives a good indication of the copy number distribution its quantitative results rely on correlating the marker protein production and fluorescence to plasmid copy number. This protocol involves many assumptions on the linearity of protein expression and fluorescence response. Also, the method only verifies one region of the plasmid and therefore does not provide a measure of plasmid structural stability. The issue of structural stability can in part be addressed by using agarose gel electrophoresis or HPLC to determine the different plasmid sizes in the sample which would directly indicate structural instability; however, the sensitivity of these methods is still limited. To increase the sensitivity of structural stability one could digest the plasmid and analyze the linear regions; however, this necessitates very pure plasmid DNA.

The different methods of determining plasmid stability suffer limitations that require a combination of these methods so that the effect on product formation can be quantitatively determined and appropriate cause and effect relationships determined. An alternative to a combination of measurements is to adopt the fourth method which is a very practical definition of genetic stability. This method follows product protein production levels through a series of serially diluted shake flasks that do not have selective pressure. This is a particularly convenient method for studying filamentous microorganisms like *Streptomyces* whose morphology prohibits plating/replica plating as a reliable method of measuring genetic stability. An example of results from this method is shown in Figure 2 (15). A recombinant *Streptomyces lividans* was transformed with a pIJ702 derivative plasmid conferring thiostrepton resistance and expression of parathion hydrolase (a heterologous protein). The culture was grown for 24 hours in LB medium in the absence of thiostrepton and the concentration of parathion hydrolase measured. Subsequently, this culture was used to inoculate another nonselective medium which in turn was grown for 24 hours, then sampled.

This process was repeated to identify any loss in production resulting from genetic instability. As Figure 2 shows, stability problems were not significant over 20 generations without selective pressure. Although convenient and practical, this method necessitates great care be taken to reproducibly prepare the media and inocula used through this experiment in order to remove the influence of environmental effects from the generation effects.

Two other effects that influence the observed instability is variation among colonies and the dependence on growth rate. To illustrate these phenomena, the results of Coppella and Dhurjati (16,17) are informative. The genetic stability of the full 2 μm based plasmid pYαEGF-25 in a cir⁰ host was examined by following the % recombinant cells through plating on nonselective agar and replica plating onto selective media using leucine selection. Studies were conducted on two different cultures that were transformed with the same vector using the same host, AB103.1. The first culture tested retained 100% recombinant cells after growth on 10 g/l glucose and the other culture retained less than 85% recombinant cells. Furthermore, the loss in the fraction of recombinant cells showed a strong dependence on the growth rate of the yeast. This loss was attributed to the highly asymmetric division of yeast cells during high growth rate (18) resulting in unequal partitioning of the plasmid copies.

NONGENETIC PROBLEMS

The objective of this section is very simple: to examine nongenetic problems that effect production of a desired recombinant protein. The significance of this objective is: (1) these factors are often overlooked or put off to a time later in the development of a particular project; and (2) by overlooking these effects (particularly early on) the optimum colony or transformant can be discarded because it was evaluated under non-optimum conditions. We have further categorized nongenetic problems into those related to the cell physiology and those related to product degradation as both affect the observed productivity.

Nongenetic problems can affect the product formation by altering the cell metabolism, thereby influencing growth and the direction of catabolism and thus product formation kinetics and yield. Two effects on production not considered in this section are those of physiological conditions on plasmid maintenance or the kinetics of transcription, translation, or secretion which are governed by the promoter constructions. We will examine four nongenetic problems that influence product formation and examine specific examples where unexpected losses in production resulted from environmental changes.

MEDIUM EFFECTS.

We will only address medium effects on systems that *a priori* were not known to be inducible or repressible as opposed to such systems as β-galactosidase which is a classic example of a well-characterized inducible operon. Let us examine four different categories of medium effects: (1) effect of selection, (2) minimal vs. rich media; (3) primary carbohydrate source (e.g. LB vs. glucose+LB); and (4) buffering additives.

Minimal and rich media can have profoundly different effects on the cell growth and the production of specific proteins. Coppella and Dhurjati (17) investigated the effect of different media on production of human epidermal growth factor (hEGF) from recombinant yeast. The media were screened by inoculating shake flasks containing 10 g/l glucose and then determining ethanol and hEGF at 48 and 72 hours. The ethanol was used as a measure of cell growth, since in this set of experiments all the glucose was fermented to ethanol within 20 hours with a yield of 0.18 g dry cell weight/g glucose consumed and the ethanol subsequently consumed with a constant yield of 0.53 g dry cell weight/g ethanol consumed.

Four different media were screened, all containing 10 g/l glucose as the primary carbon source: Yeast Nitrogen Base (YNB) media without leucine (1) and with leucine (2), YNB without leucine but with Casamino Acids added (3), and finally Yeast Extract Peptone and Dextrose with leucine (4). The effect of selection can be seen in Table I below by comparing (1) and (2). By using selective pressure (LEU⁻) growth was significantly slowed as seen by the high residual ethanol concentration and a correspondingly lower hEGF concentration which is produced in association with growth. The effect of richer media can be seen by comparing (2), (3) and (4). The richer media (3) and (4) both had significantly higher hEGF concentrations after 48 hours due in part to the faster growth. The richer media (3) & (4) also had a higher yield of hEGF per g of ethanol consumed than did the minimal media (2). However with media (3) most of the hEGF was degraded by 72 hours while the hEGF concentration continued to increase with media (4). Currently we do not understand the cause of hEGF degradation with media (3). The conclusion here is that conditions which are expected to select for the recombinant (leu- conditions) were poorest for production of hEGF and that media can affect cell growth and protein production quite differently.

Table I. The effect of media on the cell growth and production of human epidermal growth factor from recombinant yeast (adapted from (17))

media	48 hours		72 hours	
	ETHANOL (g/l)	hEGF (mg/l)	ETHANOL (g/l)	hEGF (mg/l)
(1) YNB minimal LEU⁻	2.8	0.6	1.0	1.8
(2) YNB minimal LEU⁺	2.2	0.9	0	3.0
(3) YNB rich w/ CAA	2.0	7.1	0	0.6
(4) YEPD rich LEU⁺	0.2	10	0	13

Another example of richer media significantly improving production but not cell growth is the study of parathion hydrolase production from recombinant *Streptomyces lividans* pRYE1 (19). Seven different defined media taken from the literature were compared to LB. The dry cell concentrations for the defined media were comparable to that on LB; however, the final concentration of parathion hydrolase in all the defined media was less than 10% of that with LB. Again this illustrates the uncoupling of media affects on cell growth and recombinant protein production.

Further studies examined the effect of glucose addition to LB on cell growth, and parathion hydrolase production and secretion (Figure 3). At low levels, we believe glucose provides a more readily utilizable carbon and energy source which results in improved growth and production. However at glucose levels greater than 30 g/l the level of nonoxidative metabolism (e.g. acid production) increased, as seen from the declining pH. Under these conditions growth and enzyme production decreased.

Although both cell mass and parathion hydrolase had an optimum glucose concentration of 30 g/l, the % extracellular enzyme activity of total activity was optimum for a glucose concentration of 10 g/l as shown by Table II. The observation that glucose levels affect secretion efficiency (higher % extracellular activity) is an interesting one and is not understood at this time. The effect of high levels of glucose on secretion was also seen in fermentations with 30 g/l glucose, where batch runs resulted in only 80 - 85 % of the enzyme being secreted and fed batch runs resulted in 100% of the enzyme being secreted.

Table II.   The effect of glucose concentration in LB on the secretion efficiency of parathion hydrolase from *Streptomyces lividans* pRYE1 (adapted from (19))

initial glucose concentration (g/l)	0	10	20	30	40
% extracellular parathion hydrolase activity	70	100	95	85	85

Quite often buffers are added in shaken flask experiments to control pH. However DelaCruz (19) found two popular buffers to significantly decrease productivity. Figure 4 shows results for shake flasks with recombinant *Streptomyces* expressing parathion hydrolase in LB media with phosphate and MOPS buffers. The top graph shows that increasing phosphate concentrations improved pH control, and resulted in increased growth. However the enzyme activity was substantially decreased. In the bottom graph, it can be seen that enzyme activity increases with MOPS concentration but then decreases with larger concentrations. Cell mass also increased initially with MOPS but plateaus soon after.

Thus, with the above examples we have seen that there may be an uncoupling of the effects of media on cell growth and recombinant protein production and that the media can affect the heterologous protein:
(1) production rate,

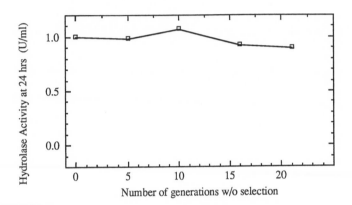

Figure 2. Parathion hydrolase activity determined after 24 hours of
incubation in nonselective media in serially diluted shake flasks to
measure the effects of genetic instability. (adapted from (15))

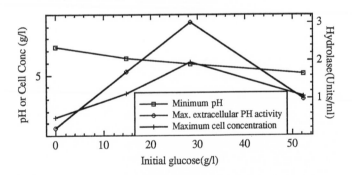

Figure 3. Minimum pH, maximum parathion hydrolase activity, and
maximum dry cell concentration as a function of initial glucose added to
LB for *Streptomyces lividans* pRYE1. (adapted from (19))

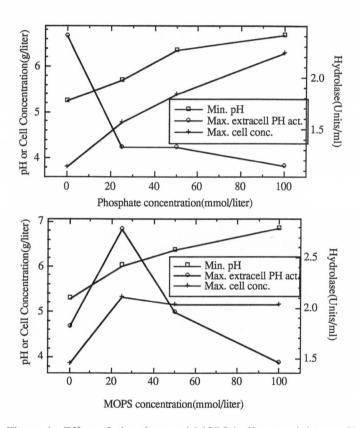

Figure 4. Effect of phosphate and MOPS buffer on minimum pH, maximum cell concentration, and maximum parathion hydrolase concentration in shaken flask experiments using LB and *Streptomyces lividans* pRYE1. (adapted from (19))

(2) product yield,
(3) secretion efficiency and
(4) product degradation.

It is likely that many, if not most, of these differences resulted from physiological rather than plasmid instability.

NONOXIDATIVE PRODUCTS UNDER AEROBIC CONDITIONS.

This discussion will include the redirection of cell metabolism to produce nonoxidative products despite aerobic conditions. This change in metabolism can arise either through catabolite control or the introduction of metabolic stress such as overproduction of recombinant protein. Note that we are limiting the discussion to growth in an aerobic environment.

A classic example of catabolite control is the Crabtree Effect in yeast where despite the availability of oxygen, glucose over a concentration of 0.1 g/l is fermented to ethanol which is subsequently oxidized after glucose consumption. This change in the metabolism results in a drop in the cell mass yield on glucose from 0.5 g/g for glucose oxidation to 0.18 g/g for glucose fermentation and an increase in the maximum specific growth rate from 0.25 hr-1 for glucose oxidation to 0.44 hr-1 for glucose fermentation (16). As another example, Snay et al. (20) found production of acetic acid, formic acid, lactic acid, and acetoin from *B. subtilis* to be induced by high dilution rates or high glucose feed levels for chemostat cultures. In addition to a change in acid production, a change in cell yield was also observed. Robbins and Taylor (21) observed nearly stoichiometric production of acetate from growth of recombinant *E. coli* on glucose which was subsequently consumed following glucose exhaustion. After controlling the glucose addition rate to minimize nonoxidative products, the cell mass yields increased significantly. Thus for many species, nonoxidative products can be observed despite aerobic conditions and these products can strongly affect cell growth rates and cell mass yields. The high level of nonoxidative product formation observed by Robbins and Taylor is quite extreme for *E. coli* however it does exemplify a good point. Production of nonoxidative products under aerobic conditions is well documented with yeast but not necessarily with other microorganisms such as *E. coli*. In fact most published data involving growth studies of *E. coli* rarely publish measurements of nonoxidative products and as can be seen by the results of Robbins and Taylor, this omission of measurement could lead to erroneous interpretations of cell yields, growth rate, and physiology.

OXYGEN TRANSFER LIMITATIONS.

Oxygen limitations have been shown in the literature to produce three types of effects in recombinant microorganisms: (1) change from oxidative to fermentative catabolism (Pasteur Effect); (2) loss in genetic stability; and (3) loss in production and/or increase in degradation. When the oxygen supply is removed from a yeast culture aerobically growing on glucose, the catabolism quickly shifts to anaerobic (the Pasteur effect) which affects both cell growth and substrate yields and is discussed in many biochemistry texts (22). A similar phenomenon occurs in *E. coli* but is not as pronounced and is not as widely known. Fass et al. (23) found that cell yield and specific growth rate of *E. coli* ATCC 25290 grown on glucose minimal media dropped when dissolved oxygen became limiting late in the fermentation. By supplementing the air feed with oxygen they were able to maintain a high growth rate and cell yield. Hopkins et. al. (24) had studied the effect of oxygen starvation on recombinant *E. coli*. They found that pKN401 was stable in *E. coli* AB1157 under controlled aerobic conditions. However after shutting off the air supply and depleting the dissolved oxygen, the fraction of recombinant cells decreased.

The third type of effect of oxygen starvation is shown in Figure 5 for *Streptomyces lividans* pRYE1 (19). Oxygen starvation preceded the onset of loss in a recombinant product (parathion hydrolase) during a fed batch fermentation. Not only does the oxygen starvation affect the recombinant protein but also the cells' metabolism as shown by the carbon dioxide production rate (CPR) and culture fluorescence. The CPR is a direct measure of metabolism while culture fluorescence is at excitation/emission wavelengths of 340 and 460 nm and indicates intracellular NAD(P)H levels. Figure 5 shows that both the CPR and culture fluorescence dropped sharply at the time of oxygen starvation. Thus oxygen transfer limitation has been shown to affect the plasmid stability, cell physiology, and recombinant protein production.

PRODUCT DEGRADATION.

Loss in product concentration or activity results from the rate of degradation exceeding the rate of product synthesis which results from a combination of an increase in the degradation rate or a decrease in the synthesis rate. Therefore it is very important for both production as well as application of an enzyme product to be able to describe the degradation behavior. This protein degradation can result from general or specific proteolysis or from inactivation due to chemical instability. Chemical instability could be further categorized into reversible and non-reversible inactivation. An example of reversible instability would be inclusion body formation from which some enzyme can be resolubilized and reactivated. Irreversible inactivation would include chemical bond changes such as oxidation and hydrolysis of the enzyme.

We have seen two good examples of product degradation thus far. The first example was shown in Table I when, during growth on YNB w/ CAA the hEGF had degraded from 7.1 mg/l at 48 hrs. to 0.6 mg/l at 72 hrs, whereas no degradation was observable with the other media tested. In this case the product loss must be due to proteolysis since the hEGF was stable with the other media (data not shown); however, as discussed before, the mechanism is not yet understood. The second example is illustrated in Figure 5 when the parathion hydrolase activity peaked at 90 hrs and was nearly all degraded by 115 hrs. This loss in activity is a combination of chemical instability and specific proteolysis as the measured chemical instability alone cannot account for the product loss and general protease activity is low throughout the fermentation.

Degradation rates due to inactivation (chemical instability) can have two obvious environmental dependencies: temperature and pH. To illustrate this dependency, the results of DelaCruz (15,25) will again be used. The loss in parathion hydrolase concentration during the fermentation process indicated a significant change in the parathion hydrolase degradation rate without appreciable general proteolysis levels. To quantify the chemical instability, cell free samples of the enzyme in media

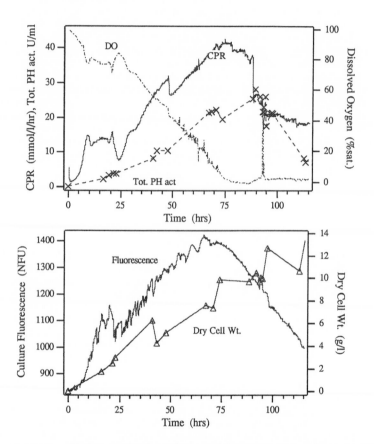

Figure 5. Carbon dioxide production rate, parathion hydrolase activity, dissolved oxygen, culture fluorescence, and dry cell concentration during a fed batch fermentation (glucose + tryptone) of *Streptomyces lividans* pRYE1. (adapted from (*19*))

were first incubated at a pH of 7.0 at different temperatures and the activity determined under standard conditions with time. The semilogarithmic plot of activity vs. time shown in Figure 6 suggested inactivation was first order with respect to the enzyme activity. The instability of the enzyme is seen to increase dramatically with increasing temperature.

A similar first order decay in enzyme was also observed in our studies on pH effects on parathion hydrolase. Again we see a large change in the chemical stability of the enzyme with in this case a change in the pH.

A word of caution. Although the inactivation kinetics' dependence on enzyme concentration in both the temperature and pH studies were found to be linear. This is not always the case, particularly if more than one mechanism is contributing to the inactivation. Therefore one should never use single point determinations for degradation rates unless sufficient evidence is available to justify such an assumption. Also since inactivation rates can be significant, samples taken from production should be analyze before and after incubation without cells to measure protein degradation rates before a synthesis rate can be determined.

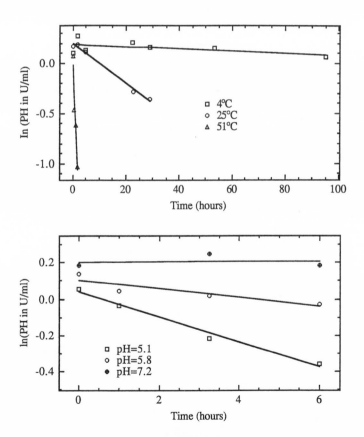

Figure 6. Temperature and pH dependence of parathion hydrolase inactivation in a cell-free system. Enzyme was produced from recombinant *Streptomyces lividans* pRYE1. (adapted from (19))

CONCLUSIONS

It is generally agreed that genetic constructions and instabilities can significantly affect the expression of heterologous proteins. Although genetic issues cannot be ignored, we have presented several examples which we believe illustrate the importance of nongenetic issues. To illustrate the importance of culture physiology, we cite examples where: considerable changes in productivity resulted from media modifications; byproduct acids and alcohols accumulated during cultivation, even under aerobic conditions; and oxygen limitations greatly reduced productivity. Also two examples were cited in which pre-formed foreign proteins were observed to be rapidly degraded in the late stages of fermentation. Although these nongenetic issues are obviously important in process optimization, they may play an important - and unrecognized - role in the selection of clones for further study. For instance, we have observed that several media which support reasonable growth have resulted in poor production by a clone which has otherwise performed very well. If these media had been used for screening, this clone may have been eliminated from our study. Also, if a product is susceptible to rapid inactivation, then the characterization of a clone based on a single time measurement of product concentration is susceptible to considerable error.

ACKNOWLEDGMENTS

We wish to acknowledge the assistance from our collaborators Dr.'s B. M. Pogell, M. K. Speedie and J. S. Steiert. Laboratory support was provided by the University of Maryland's Engineering Research Center. Financial support was provided by the University of Maryland's Center for Agricultural Biotechnology and Medical Biotechnology Center of the Maryland Biotechnology Institute. We also wish to thank New Brunswick Scientific for the use of fermentation equipment.

LITERATURE CITED

1.   Summers, D. K. ; Sherratt, D. J., Cell, 36, 1097 (1984).
2.   Komatsubara, S., Taniguchi, T. ; Kisumi, M., J. Biotechnol., 3, 281 (1986).
3.   Nishimura, N. Komatsubara, S. ; Kisumi, M., Appl. Environ. Microbiol., 53, 2800
4.   Seo, J. ; Bailey, J. E., Biotechnol. Bioeng., 30, 297 (1987).
5.   Koizumi, J. ; Aiba, S., Biotechnol. Bioeng., 28, 311 (1986).
6.   Seo, J. ; Bailey, J. E., Biotechnol. Bioeng., 28, 1590 (1986).
7.   Larsen, J. E. L.; Gerdes, K., Light, J. ; Molin, S., Gene, 968, 45 (1984).
8.   Nugent, M. E.; Primrose, S. B. ; Tacon, W. C. A., Developments in Industrial Microbiology, 21, 271 (1983).
9.   Futcher, A. B. ; Cox, B. S., J. Bacteriol., 157, 283 (1984).
10. Coppella, S. J., PhD Thesis, University of Delaware (1987).
11. Parker, C. ; DiBiasio, D., Biotechnol. Bioeng., 29, 215 (1987).
12. Walmsley, R. M.; Gardner, D. C. J.; Oliver, S. G., Mol. Gen. Genet., 192, 361 - 365 (1983).
13. Zhu, J.; Contreras, R.; Gheysen, D.; Ernst, J. ; Fiers, W., Bio/Technol., 3, 451 (1985).
14. Hollenberg, C. P., Curr. Top. Microbiol. Immun., 96, 119-144 (1982).
15. Payne, G. F.; DelaCruz, N.; Coppella, S. J., Applied Micro. Biotech., in press (1990).
16. Coppella, S. J. ; Dhurjati, P., Bioproc. Eng., 4, 75 - 80 (1989).

17. Coppella, S. J. ; Dhurjati, P., Biotechnol. Bioeng., 33, 976 - 983 (1989).
18. Coppella, S. J. ; Dhurjati, P., Biotechnol. Bioeng., in press (1990).
19. DelaCruz, N., MS Thesis, University of Maryland Baltimore County, Baltimore, MD 21228
20. Snay, J.; Jeong, J. W.; Ataai, M. M., Biotechnology Progress, 5, 63 - 69 (1989).
21. Robbins, J. W.; Taylor, K. B., Biotechnol. Bioeng., 34, 1289 - 1294 (1989).
22. Zubay, G., Biochemistry, Addison-Wesley Publishing Co., Reading Massachusetts (1984).
23. Fass, R., Clem, T. R. and Shiloach, J., *Applied Environ. Microbiol.*, 55, 1305 - 1307 (1989).
24. Hopkins, D. J.; Betenbaugh, M. J.; Dhurjati, P.; Biotechnol. Bioeng., 29, 85 - 91 (1987).
25. Coppella, S. J.; DelaCruz, N.; Payne, G. F.; Pogell, B. M.; Speedie, M. K.; Karns, J. S., Sybert; E. M.; Connors, M. A., *Biotechnology Progress*, in press (1990).

RECEIVED June 4, 1991

Chapter 2

The Epsilon Translational Enhancer

Application for Efficient Expression of Foreign Genes in *Escherichia coli*

Peter O. Olins, Catherine S. Devine, and Shaukat H. Rangwala

Monsanto Corporate Research, Monsanto Company, 700 Chesterfield Village Parkway, St. Louis, MO 63198

Escherichia coli is often the host of choice for recombinant protein production. It is particularly valuable when the cost of production is important, and a homogeneous protein is required. In addition, a major advantage of expression in *E. coli* over mammalian cells is that scale-up from a shake-flask experiment to a large-scale process is generally a fairly rapid and reliable procedure. Although expression levels *per se* may not always be an issue for high "value-added" pharmaceutical products, efficient expression is often a key factor in achieving a simple and consistent commercial process. In addition, it is desirable to have a reliable route for producing high levels of recombinant protein for research.

Although a variety of expression vectors and strategies have been described, there is still a need for an expression vector which gives useful levels of protein production with a wide range of foreign genes. In this report we describe the development of an expression vector which combines a strong, inducible promoter (the *recA* promoter of *E. coli*) and a highly efficient ribosome binding site (the "*g10*-L" sequence, based on a sequence from T7 bacteriophage). This vector was successfully used for the expression of genes from a variety of sources.

In order to determine the mechanism whereby the *g10*-L sequence gave such efficient translation initiation, a series of variant ribosome binding sites was constructed. A short sequence within the mRNA was found to be primarily responsible for the effect (denoted "Epsilon"), and the Epsilon translational enhancer was found to function even when placed at different parts of the mRNA.

Multiple factors determine the efficiency of translation initiation in prokaryotes

The prokaryotic ribosome associates in close proximity with a stretch of mRNA spanning about 40 nucleotides (1). In principle, therefore, any sequence element within this region could have an impact on the initiation process. Although analysis of natural ribosome binding site (RBS) sequences has revealed few absolute requirements, apart from the conserved "Shine-Dalgarno" (SD) sequence (2), it is clear that the region of mRNA surrounding the initiator codon does exhibit a biased base-composition (3,4). In particular, it appears that there is a bias against G and C residues in the RBS region. Although the role of this sequence bias has not been demonstrated in natural RBS regions, evidence is accumulating using foreign coding regions that the presence of a high A+T composition is associated with efficient translation initiation (reviewed in 5,6). This may be due to a direct effect of the mRNA sequence (through a sequence- specific interaction with the ribosome), or it may be an indirect effect (as a consequence of the formation of mRNA secondary structures which reduce translation efficiency). It should be pointed out, however, that the initiation process is a very complex event, and with a wide variety of recombinant proteins it has been our experience that there is no simple correlation between base- composition and expression level.

Absence of secondary structure is insufficient to explain the differences in translation initiation observed

During studies on translation efficiency in our laboratory, using a wide variety of natural E. coli genes such as aroA or cat, we found that a wide range of synthetic RBS sequences had only a small effect on expression level, and that the differences observed could not be explained by changes in predicted mRNA secondary structure (Galluppi, G. R.; Rangwala, S. H.; Olins, PO., unpublished). At least three interpretations of these results are possible: either secondary structure prediction algorithms are unreliable, small changes in structure may result in large effects on translation, or secondary structure is only a minor element in determining overall translation efficiency. In general, it appeared that the expression of several natural *E. coli* genes was relatively insensitive to the untranslated sequences placed upstream of the initiator codon, and that it was fairly easy to obtain a moderate to high level of expression of native genes. In contrast, it has been our experience that the expression of foreign genes is very variable, and is highly sensitive to the untranslated mRNA sequence used. One simple interpretation for these results is that the coding sequences of bacterial genes may already contain information important for translation initiation, and that the only further requirement in the untranslated portion of the mRNA is a correctly- positioned SD sequence. Consequently, since foreign genes probably have suboptimal sequences early within their coding regions, efficient translation may be far more dependent on the mRNA sequence upstream of the initiator codon. Other researchers have sought to optimize the mRNA sequence within the coding region of foreign genes by incorporating codons with a high A+T content in the degenerate position (7). However, this strategy is only effective in some cases (Easton, A. M.; Braford-Goldberg, S.; Devine, C. S.; Olins, P. O., unpublished), and is limited by the choice of codons available for a given coding region.

Translation efficiency a key factor in expression of foreign genes

Various strategies are possible for achieving high-level expression of foreign genes in *E. coli*. These include increasing gene dosage by using a high copy number plasmid, using strong promoters, increasing mRNA stability, improving translation efficiency and reducing proteolytic degradation. In our experience all these approaches can influence the final level of recombinant protein accumulation to some degree, but translation initiation is the most common limiting factor. In an attempt to overcome the problems of foreign gene expression in a more generic fashion, we have sought to identify natural prokaryotic RBS sequences which have a high probability of being effective for efficient translation in *E. coli*. The untranslated upstream sequence from gene *10* of bacteriophage T7 was selected as a candidate, since it is known that gene *10* is highly expressed after phage infection, and the mRNA is particularly stable (8). The "gene *10* leader" or " *g10*-L" is a synthetic RBS based on this region, and includes additional sequences which make it a convenient expression cassette (see Figure 1 and ref. 6). The expression vector used in this work is illustrated in Figure 2. It is based on plasmid pBR327, and carries an ampicillin resistance gene as a selectible marker. The plasmid includes the *recA* promoter, which can be readily induced by the addition of nalidixic acid (6,9), the *g10*-L RBS, and convenient *Nco* I and *Hind* III restriction sites for insertion of coding regions of interest. Although the *recA* promoter of *E. coli* has not been widely used for regulated expression, it offers a number of features which make it attractive for recombinant protein expression: it is well regulated at high plasmid copy number, can be strongly induced by the addition of nalidixic acid, and functions well in most *E. coli* hosts. In our laboratory the *recA* promoter has been used successfully for the expression of over 100 different genes in *E. coli*.

The *g10*-L RBS was placed upstream of a wide variety of genes from prokaryotic and eukaryotic sources: in all cases it was found to result in superior gene expression, when compared with a synthetic "consensus" RBS sequence (6). As shown in Table I, the increase in expression was often 40-fold or more. Moreover, the *g10*-L RBS is particularly effective for directing expression of foreign genes (6; Devine, C. S.; Braford-Goldberg, S.; Olins, P. O., unpublished).

The effectiveness of the *g10*-L RBS is primarily due to efficient translation initiation.

Two approaches were taken in order to determine the mechanism of this enhanced expression. In the first, a dicistronic plasmid construction was made in order to study the effect of the presence of the *g10*-L RBS on the expression of distal genes (6). In a more direct experiment, translation efficiency was monitored by comparing steady-state mRNA and protein levels (10). In both cases, it was concluded that the efficiency of the *g10*-L RBS is mainly due to translation initiation, although there was also some effect on mRNA stability.

The *g10*-L RBS contains a novel translational enhancer element, "Epsilon"

Comparison of the *g10*-L RBS with known prokaryotic ribosome binding sites showed surprisingly few sequence similarities. However, there is a sequence similarity between the *g10*-L RBS and the complement of the *E. coli* 16S rRNA, suggesting that the *g10*-L RBS might be able to form a novel, stable interaction with the 16S rRNA (10). To test this hypothesis, the short sequence (denoted "Epsilon") was inserted into otherwise poor RBS sequences, and the effect on translation efficiency was measured using *lacZ* as an assayable gene. A dramatic 100-fold increase in translation efficiency was observed when a short (9-base) Epsilon sequence was inserted upstream of the SD sequence in the mRNA. Remarkably, the effect was also observed when an Epsilon sequence was inserted downstream of the initiator codon, indicating that the sequence is acting as a translational "enhancer". A thorough analysis of published prokaryotic RBS

* * * * * * * *

UCUAGAAAUAAUUUUGUUUAACUUUAAG<u>AAGGAG</u>AUAUAUCC<u>AUG</u>

Figure 1.   Key segment of the g10-L RBS.   The initiator codon
and SD sequences are underlined, and the Epsilon translational
enhancer is denoted by asterisks.

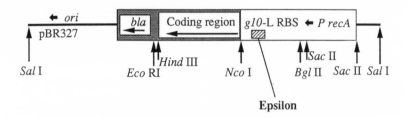

Figure 2.   Versatile expression vector incorporating the *recA*
promoter, *g10*-L RBS and Epsilon translational enhancer.

Table I. Efficiency of gene expression obtained using the *g10*-L RBS compared
to that obtained with a consensus RBS

Coding region	Fold increase in expression
E. coli lacZ	100
E. coli cat	1.6
E. coli aroA	12
Agrobact. tumefaciens IPT 1	340
Maize GST 1	53
Rat atriopeptigen	40

regions (10; unpublished observations) revealed the presence of Epsilon sequences in only a small subset of natural RBS's: in general,
there was a correlation between the presence of an Epsilon sequence
and highly-expressed genes.   Examples of this include a variety of
structural proteins from bacteriophages, and the *lpp* gene from *E. coli*,
which is the most abundant protein in the cell.

Future perspectives

It seems possible that the mRNAs from other other organisms may have evolved to interact with the 16S ribosomal RNA in a similar fashion to the Epsilon sequence of T7 bacteriophage, and it may therefore be valuable to examine mRNA sequences from organisms other than *E. coli*. In addition, since the structure of ribosomes has been conserved over such a long evolutionary period (11,12), it is conceivable that it might be possible to engineer the mRNA sequences of organisms other than *E. coli*, such that they could form a stable interaction with the cognate small ribosomal RNA, and hence obtain an increase in translation efficiency.

References

1. Argetsinger Steitz, J. A. *Methods Ezymol.* 1979, *60*, 311-321.

2. Shine, J.; Dalgarno, L. *Proc. Natl. Acad. Sci.* USA 1974, *71*, 1342-1346.

3. Gold, L.; Pribnow, D.; Schneider, T.; Shinedling, S.; Singer, B. S.; Stormo, G. *Ann. Rev. Microbiol.* 1981, *35*, 365-403.

4. Gren, E. J. *Biochimie* 1984, *66*, 1-29.

5. Stormo, G. D. In *Maximizing Gene Expression;* Reznikoff, W.; Gold. L., Eds.; Butterworth: USA, 1986, pp. 195-224.

6. Olins, P. O.; Devine, C. S.; Rangwala, S. H.; Kavka, K. S. *Gene* 1988, *73*, 227-235.

7. Hsiung, H. M.; MacKellar, W. C. *Methods Ezymol.* 1987, *153*, 390-401.

8. Dunn, J. J.; Studier, F. W. *J. Mol. Biol.* 1983, *166*, 477-535.

9. Feinstein, S. I.; Chernajovsky, Y.; Chen.; Maroteaux, L.; Mory, Y. *Nucleic Acids Res.* 1983, *11*, 2927-2941.

10. Olins, P. O.; Rangwala, S. H.; *J. Biol. Chem.* 1989, *264*, 16973-16976.

11. Gutell, R. R.; Weiser, B.; Woese, C. R.; Noller, H. F. *Prog. Nucleic Acid Res. & Mol. Biol.* 1986, *32*, 155-216.

12. Huysmans, E.; de Wachter, R. *Nucleic Acids Res.* 1986, *14*, 73-118.

RECEIVED June 7, 1991

Chapter 3

High-Cell-Density Fermentations Based on Culture Fluorescence

Applications for Recombinant DNA Products in *Escherichia coli*

S. A. Rosenfeld[1,3], J. W. Brandis[1,4], D. F. Ditullio[1,5],
J. F. Lee[2,6], and W. B. Armiger[2,6]

[1]Department of Molecular Biology, Triton Biosciences Inc.,
Alameda, CA 94501
[2]BioChem Technology, Inc., Malvern, PA 19355

A method was developed that uses on-line culture fluorescence to optimize fermentor productivity of an Escherichia coli strain encoding human transforming growth factor-alpha (TGF-α). Typical batch mode fermentations yielded 10-12 g/l cell dry weight. Higher cell densities (40 g/l) were achieved via fed batch fermentations using culture fluorescence to measure on-line cell mass and automatically determine glucose addition rates. Under these conditions, commercially viable amounts of TGF-α were obtained and accumulated as refractile inclusion granules inside the cells. This computer driven on-line fluorescence measurement provides an efficient and reproducible means of monitoring fermentation behavior. Culture fluorescence methodology may be used with any strain of E. coli, and any size fermentation vessel.

Commercialization of recombinant proteins produced by *E. coli* require developing fermentation methodologies that yield high and efficient product formation. Some of the requirements necessary for a well-defined and successful fermentation may be summarized in Table I.

[3]Current address: Dupont Merck, 500 South Ridgeway Avenue, Glenolden, PA 19036
[4]Current address: RiboGene, Inc., Dublin, CA 94568
[5]Current address: Berlex Biosciences, Inc., Alameda, CA 94501
[6]Current address: BioChem Technology Inc., King of Prussia, PA 19406

Table I. Fermentation Requirements

Fermentation Monitoring System - On Line
Fermentation Controlling System - Automatic
Data Acquisition, Storage and Manipulation
Versatility
Research Tool

It is desirable to develop fermentations that are easily monitored, controlled, and provide a research tool for understanding "fermentation physiology". The FERMAC (Fermentation Monitoring, Analysis and Control System) software developed by BioChem Technology Inc. fulfills these requirements.

The basis of the control system is an on-line measurement of culture fluorescence generated from the reduced form of intracellular nicotinamide adenine dinucleotides (NADH and NADPH or more simply NAD[P]H). All living cells contain these cofactors which are essential for numerous metabolic reactions. The reduced electron carrier, NAD[P]H, fluoresces at 460 nm when irradiated with light at 340 nm, whereas the oxidized molecules, NAD[P], do not fluoresce.

The culture fluorescence signal is a composite of the intracellular NAD[P]H pools, environmental effects (ie. temperature, pH, fermentor agitation, etc.) and metabolic effects (biochemical pathways used by the cells during different growth conditions). Eliminating metabolic and environmental effects by maintaining the same fermentation conditions and growth medium from run to run results in a culture fluorescence dependent on intracellular NAD[P] pools only. Since the reduced nucleotide pools remain relatively constant under a given metabolic condition, culture fluorescence will be a function of cell concentration only (Figure 1). The consequences of an on-line determination of cell concentration permits a carefully controlled glucose feed regimen that achieves high dry cell density fermentations that are highly reproducible and amenable to automatic computer control. This system has been successfully applied to the production of *E. coli* derived recombinant TGF-α at commercially acceptable levels.

Methods

Culture and Growth Conditions

E. coli strains encoding recombinant TGF-α were grown overnight in LB rich medium (500 ml. volume) containing ampicillin (150μ/ml) at

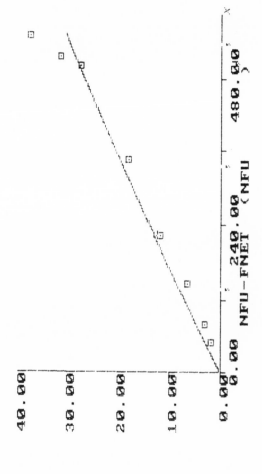

Figure 1. Correlation of fluorescence with dry cell weight during a glucose feed fermentation. The (□) symbols represent individual off-line measurements of dry cell weight. The curve is derived from on-line calculated dry cell weight as determined by the FER-MAC computer program.

32°C then added to a Braun Biostat 2E fermentor containing 8.0 liters of fermentation medium (Bacto-tryptone, 62.5 g/1; yeast extract, 31.25 g/1;K_2HPO_4, 7.47 g/1; $MgSO_4$, 1 g/1;L-serine, 0.41 g/1; L-proline, 0.44 g/1; L-arginine, 0.2 g/1; L-alanine, 1.0 g/1;L-cysteine, 0.31 g/1; glycine, 1.46 g/1; containing ampicillin (150 μg/ml).

Fermentation conditions were as follows: the impeller speed was set and held at 800 rpm; the sparging rate was maintained at 10 liters/minute using a mixture of air and bottled oxygen to keep the dissolved oxygen (DO) above 40%; pH was maintained at 7.1 with NH_4OH. Glucose was added from a 50% (w/v) reservoir as directed by a feed algorithm. The culture was grown at 32°C until such time the temperature was shifted to 42°C (during TGF-α induction experiments) automatically by the FERMAC software program (Results Section).

Analytical Methods

Culture fluorescence was measured using the FluroMeasure detector Model 50 (BioChem Technology, Inc.). FERMAC and FluroMeasure are trademarks of BioChem Technology, Inc.

Dry cell weight (DCW) measurements were performed by removing fermentation samples, centrifuging, washing the resulting pellet with phosphate buffered saline, then resuspending the pellet in distilled water. The sample was placed in tared aluminum weigh boats at 70° for 24 hours under vacuum, then weighed.

Optical density measurements of cultures were performed at A_{550}.

Glucose concentration was determined using a Yellow Spring/1 Incorporated Model 2000 glucose and 1-lactate analyzer.

TGF-α in cell extracts was determined via SDS-polyacrylamide gel electrophoresis (PAGE) using a tricine buffer system described by Schagger and Jagow(1).

Results and Discussion

High Cell Density Fermentation—Glucose Feed Studies

One goal in producing recombinant proteins in *E. coli* is achieving high cell density fermentations. Greater number of cells present per fermentation potentially yields more product per fermentation cycle. The limiting factor in typical batch fermentations is the accumulation of toxic waste products, including mixed acids, during metabolism of excess glucose (Figure 2) (2,3). In order to attain high cell density fermentations the levels of various toxic metabolic by-products, notably acetate, must be minimized. By feeding glucose during a fermentation such that its level is carefully regulated, metabolic byproducts do not accumulate and inhibit growth. Since cell densities are measured on-line throughout the fermentation cycle, it is possible to use culture fluorescence to determine glucose addition rates if the yield of cells per gram of glucose and the amount of glucose needed to maintain one gram of cells per unit time are known. The growth model that was used to control glucose feed rates is indicated on Equation I.

Equation I.   Growth Model Equation

$$\underset{\text{TERM}}{\text{GROWTH}} \qquad \underset{\text{TERM}}{\text{MAINTENANCE}}$$

$$G_N = C_T/Y_{gg} + (C_T)(Y_{gm})(DT)$$

where,

G_N = glucose needed (g)
C_T = total cells in the reactor (g)
DT = time required to double the existing total cell population (hr)
Y_{gg} = [g cells/g glucose] : growth requirement
Y_{gm} = [g glucose needed/g cell•hr] : maintenance requirement

---

On-line measurements of culture fluorescence were taken at 30 second intervals; and based on these determinations, the FERMAC computer program estimated a cell doubling time (DT) based upon previous timepoint calculation. Although the Y_{gg} and Y_{gm} terms also change throughout the fermentation, their values are set to historical values determined during steady-state growth conditions. The values of these variables in Equation I may be changed and replaced to fine tune the systems as necessary. Glucose pumping rates were modulated automatically based upon the following pump rate algorithm.

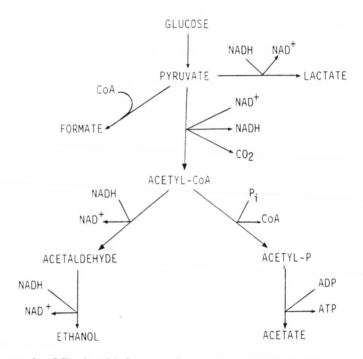

Figure 2. Mixed acid fermentation pathways utilized by *E. coli* during growth with high glucose concentrations.

Equation II. Glucose Pump Rate Algorithm

Logic Statement	Normal Pump Rate
If $C_T/G_T \geq Y_G$ and $\mu_i \geq 0.2$	Maintain
If $C_T/G_T \geq Y_G$ and $\mu_i \leq 0.2$	Increase by 5%
If $C_T/G_T \leq Y_G$ and $\mu_i \leq 0.2$	Decrease by 15%

where,
C_T = Total cells in reactor (g)
G_T = Total glucose in reactor (g)
Y_G = Average growth yield
μ_i = Instantaneous specific growth rate (hr^{-1})

The μ_i term in the pump rate algorithm is the inverse of the DT (as defined in Equation I) and Y_G is the "average" dry cell weight of cells (grams) produced per glucose (grams) consumed. The Y_G term is derived from historic data obtained with a particular cell line. All other terms in the algorithm are derived from on-line culture fluorescence measurements.

The normal pump rate (l/hr) depicted in Equation II = $(G_N/G_C)/DT$ where,
G_N = Glucose needed (as in Equation I) G_C = Glucose concentration in feed reservoir (g/l)
DT = Time required to double the existing cell population (as in Equation I).

Using this glucose feed model it was possible to achieve cell densities of about 40 g/l, 3 - 4 times the cell density of a glucose batch fermentation (Figure 1). The correlation between cell concentration and fluorescence was determined to be a 2° polynomial up to 30 g/l dry cell weight. The deviation from linearity is due to the optical characteristics of the FluroMeasure probe.

The control system software derive a calibration curve based on this relationship. Employing this calibration curve, there was a good correlation between off-line dry cell weight measurements and on-line calculated dry cell weight as determined by culture fluorescence measurements.

Application of Culture Fluorescence to TGF-α Production in *E. coli*

Having shown that the control system permitted a reliable and automatic glucose feed regimen for attaining high cell densities, the application of the system for producing recombinant DNA proteins in *E. coli* was tested.

A strain of *E. coli* that expressed recombinant transforming growth factor α (TGF-α), was grown at 32°C to an A_{550} = 30 or 275 NFU-FNET Units (a filtered fluorescence signal as defined by the FERMAC software program), followed by an automatic temperature shift to 42°C, the temperature necessary for TGF-α synthesis (Figure 3). A final temperature shift to 38°C was likewise automatically performed using FERMAC software. The glucose feed rate was monitored and controlled automatically during all stages of the fermentation. The culture attained a final A_{550} = 74, and 26 g/l DCW after 3 hrs induction before terminating the experiment. Moreover, there was good agreement between on-line and offline dry cell weight measurements (Figure 3). Glucose levels never exceeded 7.6 g/l; however finer control of residual glucose levels should be attainable by slight modifications of the parameters in Equations I and II.

TGF-α production during this glucose feed study was measured to determine if high cell density fermentation was deleterious to product formation. Results of these experiments demonstrated that cell lysates of high cell density fermentation broths contained equivalent or higher levels of TGF-α compared to low density batch fermentations (Figure 4).

A comparison of the specific fermentor productivity (g dry cell weight/liter/hr) of low cell density fermentations to the glucose feed high cell density experiment discussed above indicated that the high cell density fermentation displayed a 92% increase in fermentor productivity (Table II). Optimization of the glucose feed regimen and medium composition are being investigated and might lead to higher fermentor productivities than those reported above.

Conclusions

The present study demonstrates how culture fluorescence may be used to monitor and control *E. coli* fermentations (Table III).
Unlike other systems used to monitor and control fermentations, culture fluorescence also provides a method for analyzing cell physiology. Culture fluorescence technology gives on-line data as to the metabolic state of the cell via NAD[P]H measurements. This system may be implemented regardless of the cell type under investigation (bacterial, plant, mammalian) or the fermentor size and configuration. Future fermentation studies may exploit this information to rationally design and optimize fermentation protocols and in so doing enhance fermentor productivity.

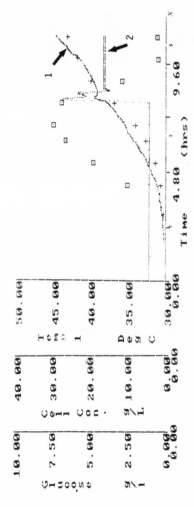

Figure 3. Temperature induction of TGF-α synthesis during high cell density glucose feed experiment. Curve 1: on-line calculated dry cell weight; Curve 2: fermentation temperature profile; (+) symbols represent off-line dry cell weight measurement; (□) symbols designate glucose levels during the fermentation.

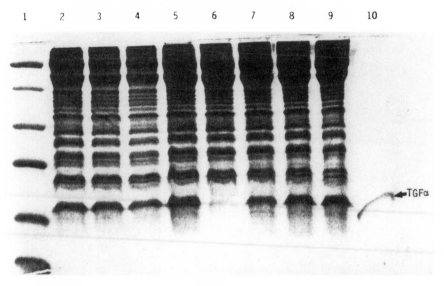

Lane	Sample
1	Low Molecular Weight Standard
2	Low Cell Density Run #89–5–1
3	Low Cell Density Run #89–5–2
4	Low Cell Density Run #TM 1399
5	High Cell Density Run #89–14
6	High Cell Density Run #89–38 T_0
7	High Cell Density Run #89–38 T_1
8	High Cell Density Run #89–38 T_2
9	High Cell Density Run #89–38 T_3
10	TGF Standard

Figure 4. SDS-PAGE analysis of TGF-α production during low and high cell density *E. coli* fermentations. Lanes 2-5 are cell lysates derived from previous low cell density and one high cell density fermentation (Run #89-14) experiments sampled 2 hours post 42°C temperature shift. Lanes 6-9 are cell lysates derived from high cell density run #89-38 prior to temperature induction, 1 hour, 1.5 hours and 2 hours post temperature shift, respectively.

Table II. Fermentor Productivity

	Low cell density	High cell density
	90g DCW/10 Liter/8 Hr	260g DCW/10 Liter/12 Hr
Specific Fermentor Productivity (g DCW/Liter/Hr)	1.13	2.17

Table III. Applications of FERMAC System

General Fermentation Monitoring (pH, agitation, dissolved oxygen, etc)
Data Acquisition, Storage, and Manipulation
Automatic Fermentation Control (glucose feed, temperature, etc)
Reproducible Fermentation Parameters
High Cell Density Fermentation - Increased Fermentor Productivity
Recombinant DNA Protein production
Versatility
Scalability - Any Vessel Size
Probing Fermentation Physiology

References

1. Schagger, H.; von Jagow, G. *Analytical Biochemistry* 1987, *166*, 368-379

2. Anderson, K.W.; Grulke, E.; Gerhardt, P., *Biotechnology.* October, 1986, 891.

3. Landwall, P.; Holm, T. *Journal General Microbiology.* 1977, *103*, 345.

RECEIVED June 26, 1991

Chapter 4

Mathematical Modeling and Optimization of Complex Biocatalysis

A Case Study of Mercuric Reduction by *Escherichia coli*

George P. Philippidis[1,3], Janet L. Schottel[2],
and Wei-Shou Hu[1]

[1]Department of Chemical Engineering and Materials Science,
University of Minnesota, Minneapolis, MN 55455
[2]Department of Biochemistry and Plant Molecular Genetics
Institute, University of Minnesota, St. Paul, MN 55108

The *mer* operon-encoded reduction of Hg^{2+} to Hg° by recombinant *Escherichia coli* cells involves a complex mechanism that consists of a transport and an enzymatic reaction step. A mathematical model developed for the transfer of Hg^{2+} across the cellular envelope by transport proteins and the subsequent reduction by the cytoplasmic mercuric reductase was supported with experimental data. The data, obtained with cells that carried the *mer* operon at various copy numbers, also helped determine the values of the parameters of the model. Transport of Hg^{2+} appears to be the rate-determining step of the overall process. Optimization of the biocatalytic rate by gene amplification was limited by the modest increase of the transport protein concentration. The model predicts that subcloning the transport genes to amplify their expression relative to that of the enzyme may lead to enhancement of the Hg^{2+} reduction rate.

The current use of industrial enzymes for bioprocessing is predominantly restricted to simple hydrolytic and equilibrium processes, such as those catalyzed by penicillin acylase, α- and β-amylases, proteases, and glucose isomerase. Advancement of the importance of biocatalysis lies in the study of complex biochemical reactions, which involve the coordinated action of several polypeptides, require provision of bioenergy and cofactors, and are coupled to the metabolism of the cell. Applications of complex biocatalysis include the production of organic chemicals, amino acids, vitamins, cofactors, and antibiotics. Such processes make whole cells, instead of purified enzymes, an attractive and economical catalyst. The inherent complexity of the cell,

[3]Current address: Solar Energy Research Institute, Biotechnology Research Branch,
Engineering and Analysis Section, 1617 Cole Boulevard, Golden, CO 80401

however, requires a new strategy for process optimization. Thus, high level gene expression, the traditional means for protein overproduction, may no longer lead to maximal productivity.

In order to optimize the rate of complex biocatalytic reactions, we need to thoroughly understand the interactions among the various components of the reaction mechanism. That can be accomplished by analyzing the process and developing a mathematical model which has the ability to describe accurately the performance of the biocatalytic system. Here we present the formulation of a deterministic model for a biocatalytic system, evaluation of its validity, and use of the model as a guide towards process optimization.

Mercuric Reduction by _Escherichia Coli_. The reduction of mercuric ions (Hg^{2+}) to elemental mercury (Hg^o) by recombinant _Escherichia coli_ has been chosen as representative of complex biocatalysis, due to its relative simplicity, adequate understanding of its structure and genetics, and its potential applicability in bioremediation of mercury-contaminated sites. Reduction of Hg^{2+} is carried out by polypeptides of the plasmid-borne _mer_ operon (1-2). The _mer_ operon of plasmid R100 encodes the periplasmic _merP_ gene product, the inner membrane _merT_ and _merC_ gene products, and the cytoplasmic _merA_ gene product (mercuric reductase) (3-4). The genes are subject to positive and negative transcriptional control exerted by the _merR_ gene product, which also negatively autoregulates its own transcription (2,5). The promoter/operator regions for the structural genes _merTPCAD_ and gene _merR_ are located between genes _merR_ and _merT_ (6-9). RNA polymerase transcribes the structural genes into a polycistronic mRNA in the order _merTPCAD_ (10-11).

Figure 1 depicts a simplified structure of the Hg^{2+} reduction system. The _merP_ and _merT_ gene products are believed to mediate the transfer of Hg^{2+} across the cellular envelope (12-13). In the absence of these proteins, the cells do not reduce Hg^{2+} (13). The participation of the _merC_ gene product in the transport of Hg^{2+} remains tentative (12). Cysteyl residues of the three polypeptides are believed to carry out the binding and transfer of Hg^{2+} (14). It has been suggested that the transport mechanism may be energy-dependent (12,15-16). In the cytoplasm, the flavoenzyme mercuric reductase catalyzes the reduction of Hg^{2+} to Hg^o using NADPH as electron donor (17). Since Hg^o is volatile, it disappears from the cell environment, thus providing a means to monitor the biocatalytic activity of the cells. A pair of cysteyl residues in the amino-terminal domain of mercuric reductase has been proposed to mediate the transfer of Hg^{2+} from the transport proteins to the active site of the enzyme (14,20). A cysteyl pair at the active site (18) is essential for Hg^{2+} reduction (19). The active site cystine and a second cysteyl pair in the carboxyl terminus are thought to be involved in Hg^{2+} binding (20-23).

According to a proposed catalytic mechanism for mercuric reductase, the enzyme (E) cycles between the four-electron reduced form $EH_2 \cdot NADPH$ and the two-electron reduced form $EH_2 \cdot NADP^+$ (23-24). Mercuric ions bind to $EH_2 \cdot NADPH$, whereas Hg^o is released from $EH_2 \cdot NADP^+$. Thiols or ethylenediaminetetraacetic acid (EDTA) are required for catalytic activity of the purified enzyme; both compounds act as effective chelating agents for Hg^{2+} (24-25). However, mercuric reductase exhibits different kinetics in the presence of the two compounds (25).

Taking the above information into account, a theoretical analysis of mercuric reduction was performed and the system was modeled. Genetic manipulations and kinetic experimentation were carried out to obtain data necessary for determination of the parameters of the model and evaluation of its validity. The model was subsequently used to investigate strategies for maximization of the biocatalytic rate of mercuric reduction.

KINETIC MODEL

Mercuric reduction at the whole-cell level involves the coordinated action of the transport proteins and mercuric reductase, which act sequentially on Hg^{2+}. The two steps, transport and reaction, were analyzed separately and a model was developed for each one. The model for the overall process is then obtained by combining the individual models.

Modeling of the Hg^{2+} Transport. A model for the Hg^{2+} transport mechanism is shown in Figure 2. Mercuric ions chelated by thiols or EDTA diffuse rapidly through the pores of the outer membrane into the periplasmic space and bind to protein P, forming the P-Hg complex. The mercuric ions are then transferred to the inner membrane protein T to form a T-Hg complex. It has been suggested that Hg^{2+} transport is energy-dependent (12,26). Accordingly, the transfer of Hg^{2+} across the inner membrane is assumed to take place at the expense of metabolic energy. Finally, the Hg^{2+} are transferred to mercuric reductase.

Assuming that steady state is rapidly established between the rates of Hg^{2+} transport through the periplasm and inner membrane, mathematical formulation of the model resulted in the following expression for the Hg^{2+} transport rate (27):

$$r_T = r_{T,max}\left(\frac{[Hg^{2+}]_o}{K_1 + [Hg^{2+}]_o} - \frac{[Hg^{2+}]_i}{K_2 + [Hg^{2+}]_i} \right) \qquad (1)$$

where:

$$r_{T,max} = (A/V)\left(\frac{[P]_t}{2/D_1} + \frac{[T]_t}{(1/D_2 + 1/D'_2)} \right) \qquad (2)$$

Equation 1 describes the transport rate r_T as a function of the extracellular, $[Hg^{2+}]_o$, and intracellular, $[Hg^{2+}]_i$, mercuric ion concentration. The parameter $r_{T,max}$ is the maximal Hg^{2+} transport rate into the cell and is a function of the concentrations of the two transport proteins P and T (Equation 2).

Modeling of the Hg^{2+} Reduction Reaction. Our model for the catalytic reaction takes into account an ordered bireactant (Hg^{2+}, NADPH) mechanism (23-24) and a general inhibition scheme (27), as shown in Figure 3. It has been assumed that mercuric ions can bind to both the free enzyme E and the intermediate form E-NADP to form abortive complexes, thus inhibiting the reaction rate. When solved, the steady state mass balance equations for the six forms of the enzyme (Figure 3b) yield the following Hg^{2+} reduction reaction rate:

$$r_R = \frac{r'_{R,max}[Hg^{2+}]_i[NADPH]_i}{\begin{array}{c}[Hg^{2+}]_i[NADPH]_i + [NADPH]_i[Hg^{2+}]_i^2/K_i' + [Hg^{2+}]_iK_{mA}(1+[Hg^{2+}]_i/K_i'') + \\ + [NADPH]_iK_{mB} + K_{mB}K_{sA}(1+[Hg^{2+}]_i/K_i')\end{array}}$$

$$(3)$$

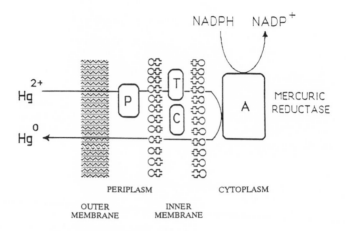

Figure 1. Simplified representation of the mercuric ion reduction system encoded by the R100 *mer* operon.

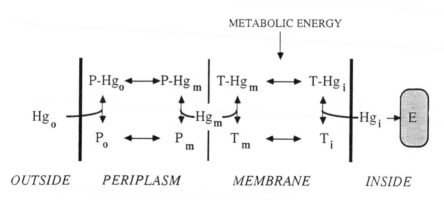

Figure 2. Schematic depiction of the Hg^{2+} transport model. P and T are the *merP* and *merT* gene products, respectively. (Reproduced with permission from reference 27. Copyright 1991 John Wiley & Sons.)

where: $r'_{R,max} = [E]_t/(1/k_5 + 1/k_7)$
$K_i' = (k_5 + k_7)k_{12}/k_5/k_{11}$
$K_i'' = k_{10}/k_9$
$K_{sA} = k_2/k_1$
$K_{mA} = k_5 k_7/(k_5 + k_7)/k_1$
$K_{mB} = (k_4 + k_5)k_7/(k_5 + k_7)/k_3$

The parameter $r'_{R,max}$ is the maximal initial reduction rate that would have been attained in the absence of substrate inhibition. It is proportional to the concentration of the mercuric reductase $[E]_t$. Equation 3 describes the dependence of the reaction rate on the intracellular concentrations of the two substrates, Hg^{2+} and NADPH, with

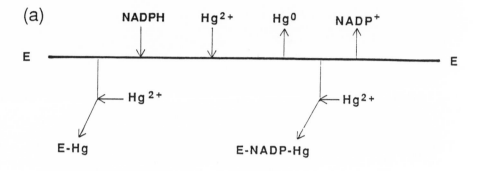

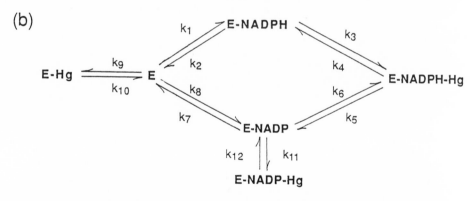

Figure 3. (a) Schematic representation of the Hg^{2+} reduction model by mercuric reductase (E). (b) Transformation of the model into a reaction scheme. (Reproduced with permission from reference 27. Copyright 1991 John Wiley & Sons.)

a decrease in the rate occurring at high Hg^{2+} concentrations. The extent of the inhibition depends largely on the value of the inhibition parameter K_i'. The relationship between the reaction rate r_R and substrate concentration can be best depicted by a 3-dimensional representation, as shown in Figure 4. The concentration of Hg^{2+} varies from 0 to 120 μM, whereas that of NADPH varies from 0 to 1000 μM.

In the presence of excess intracellular NADPH, Equation 3 simplifies to:

$$r_R = \frac{r_{R,max}\,[Hg^{2+}]_i}{K_m + [Hg^{2+}]_i + [Hg^{2+}]_i^2/K_i} \tag{4}$$

where:

$$r_{R,max} = r'_{R,max}/(1 + K_{mA}/[NADPH]_i + K_{mB}K_{sA}/[NADPH]_i/K_i'')$$
$$K_m = (K_{mB} + K_{mB}K_{sA}/[NADPH]_i)/(1 + K_{mA}/[NADPH]_i + K_{mB}K_{sA}/[NADPH]_i/K_i'')$$
$$K_i = (1 + K_{mA}/[NADPH]_i + K_{mB}K_{sA}/[NADPH]_i/K_i'')/(1/K_i' + K_{mA}/[NADPH]_i/K_i'')$$

An expression similar to Equation 3 has been proposed by Dixon and Webb for substrate inhibition in ordered enzymatic reactions (28). That rate expression also simplifies to Equation 4, when the noninhibitory substrate (NADPH in our case) is present in excess.

We have previously determined the Michaelis constant of mercuric reductase for NADPH to be 13.9 μM (29). This value is more than 10-fold smaller than the intracellular concentration of NADPH in exponentially growing *Salmonella typhimurium* (146 μM), a bacterium physiologically and genetically closely related to *E. coli* (29). Assuming that a similar NADPH concentration exists in *E. coli*, it can be considered that intracellular NADPH is indeed present in excess with regard to the requirements of the Hg^{2+} reduction reaction (Equation 3). Thus, Equation 4 can effectively represent the kinetics of the enzymatic reaction.

DETERMINATION OF THE MODEL PARAMETERS
In order to determine the values of the parameters in Equations 1 and 4, we need to measure the Hg^{2+} transport (r_T) and reaction (r_R) rates. We have previously described the use of ether-permeabilized cells to measure the activity of mercuric reductase, uncoupled from the transport system (29). In intact cells, however, the function of mercuric reductase is coupled to the action of the transport proteins. Consequently, the overall rate of Hg^{2+} reduction r_o is affected by both the transport and the enzymatic reaction rate. Assuming that steady state between transport and reduction reaction prevails, the overall rate can be considered equal to the Hg^{2+} transport rate. Under that assumption, the reduction rate of intact cells represents the activity of the Hg^{2+} transport system.

The turnover rate of Hg^{2+} by mercuric reductase was determined using ether-permeabilized *E. coli* C600 r^-m^+ cells harboring plasmid pDU1003 (29). The Hg^{2+} and NADPH concentrations were varied from 0 to 120 μM and from 0 to 1000 μM, respectively. When the measured reduction rates were fitted to Equation 3 using a nonlinear regression algorithm (Figure 5), the parameters $r'_{R,max}$, K_i', K_{sA}, K_i'', K_{mA}, and K_{mB} were found equal to 171.6 nmol Hg^{2+}/min/mg protein, 84.2 μM, 17.6 μM, 59.6 μM, 8.2 μM, and 23.3 μM, respectively. Figure 5 shows that the experimental reduction rates correlate well with Equation 3 of the model. The correlation coefficient was 0.98 at a 95% statistical confidence level. Then, the parameters $r_{R,max}$, K_m, and K_i

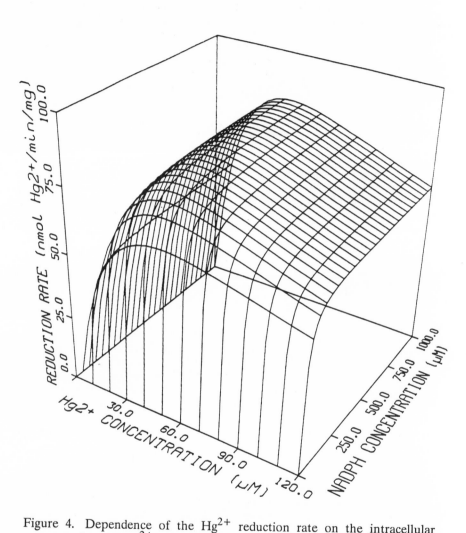

Figure 4. Dependence of the Hg^{2+} reduction rate on the intracellular concentrations of Hg^{2+} and NADPH. (Reproduced with permission from reference 29. Copyright 1990 Butterworth–Heinemann.)

of Equation 4 can be calculated using their definitions. At an NADPH concentration of 1.0 mM, $r_{R,max}$, K_m, and K_i are equal to 169.4 nmol Hg^{2+}/min/mg protein, 23.4 μM, and 60 μM, respectively.

The Hg^{2+} transport rate was measured using intact *E. coli* C600 r^-m^+ cells containing plasmid pDU1003 and the data were fitted to Equation 1. Unfortunately, the intracellular Hg^{2+} concentration of Equation 1 cannot be readily determined experimentally. However, taking into account the steady state assumption between the transport and reaction rates ($r_T = r_R = r_o$), Equation 4 can be set equal to the measured overall reduction rate r_o of intact cells and solved in terms of the intracellular Hg^{2+} concentration (27):

$$[Hg^{2+}]_i = \frac{K_i(r_{R,max}/r_o-1)-[(K_i^2(r_{R,max}/r_o-1)^2-4K_mK_i)]^{1/2}}{2} \qquad (5)$$

That expression can subsequently replace $[Hg^{2+}]_i$ in Equation 1, since $r_{R,max}$, K_m, and K_i have already been determined. Fitting the intact cell reduction rates to Equation 1 yielded for $r_{T,max}$ and K_1 the values of 17.5 nmol Hg^{2+}/min/mg protein and 3.6 μM, respectively, whereas the value of K_2 exceeded 10^6 μM. The correlation of the data to Equation 1 was satisfactory (0.96). The large value of K_2 indicates that the term of Equation 1 involving the intracellular Hg^{2+} concentration is practically negligible. Therefore, the model equation for the Hg^{2+} transport rate r_T is reduced to:

$$r_T = \frac{r_{T,max}[Hg^{2+}]_o}{K_1 + [Hg^{2+}]_o} \qquad (6)$$

From a physicochemical point of view, the high value of the dissociation constant K_2 suggests a low affinity of the inner membrane protein T for Hg^{2+} at the membrane-cytoplasm interface, thus favoring the transfer of Hg^{2+} to mercuric reductase for subsequent reduction. On the other hand, the low value of K_1 indicates a high affinity of the periplasmic protein P for Hg^{2+}, leading to efficient binding of the ions for safe transfer to protein T. The high affinity of the periplasmic component of the Hg^{2+} transport system for its substrate compares well with the reported high affinity of other transport systems, such as those of maltose and phosphates, for their substrates (K_1 values of 1 and 0.2 μM, respectively) (31).

MODEL VALIDITY EVALUATION

The validity of the model Equations 4 and 6 was tested using recombinant *E. coli* C600 r^-m^+ cells harboring the *mer* operon at five different copy numbers. Gene copy number variation is expected, in general, to result in variation of the corresponding polypeptide concentration. The construction of the recombinant plasmids has been described previously (32); their nomenclature and copy numbers, determined by dot-blot DNA hybridization, are summarized in Table I. Plasmid copy numbers varied from 3 to 140 copies per cell, representing an overall 47-fold gene amplification effect.

Intact plasmid-harboring cells were assayed to determine the Hg^{2+} transport rate at various Hg^{2+} concentrations ranging from 5 to 120 μM (32). The measured intact cell reduction rates for each plasmid construct exhibited satisfactory correlation to Equation 6 (greater than 0.92). As expected, the value of $r_{T,max}$ varied with gene copy number, reflecting its dependence on the concentration of the transport proteins

Table I. Copy numbers of the *mer* plasmids and kinetic parameters
of the Hg^{2+} transport and reduction systems in
E. coli cells harboring those plasmids [a]

PLASMID	DETERMINED COPY NUMBER (copies/cell)	$r_{T,max}$ [b]	K_1 [c]	$r_{R,max}$ [b]	K_m [c]	K_i [c]
R100	3	8.2	3.8	45	12.6	96.7
pBRmer	67	13.4	4.7	152	15.9	91.5
pDU1003	78	17.5	3.6	168	22.8	87.4
pACYCmer	124	20.6	5.1	221	15.5	75.2
pUCmer	140	19.8	6.0	305	19.5	93.1

[a] Compiled from data in Ref. 32.
[b] Expressed in nmol Hg^{2+}/min/mg protein
[c] Expressed in μM Hg^{2+}

(Table I). In contrast, the value of K_1 remained constant at about 4.6 μM, underlining the common identity of the transport proteins in all constructs (Table I).

Ether-permeabilized cells carrying the recombinant plasmids were used to determine the enzymatic reaction rate of the five plasmid constructs in the 5-120 μM range of Hg^{2+} concentrations (32). In all experiments, NADPH was present in excess (1 mM) to ensure that Hg^{2+} was the rate-limiting substrate of the enzymatic reaction. The reaction rates were then correlated to Equation 4 to examine the ability of that expression to describe the reaction rate of cells containing various concentrations of mercuric reductase. In all cases Equation 4 was able to describe satisfactorily the kinetics of the reaction with correlation coefficients exceeding 0.97. Again, the maximal reduction rate ($r_{R,max}$) increased with gene copy number, whereas the values of K_m and K_i remained essentially equal to 17.3 and 88.8 μM, respectively, confirming their identity as intrinsic parameters of mercuric reductase, independent of intracellular enzyme concentration (Table I).

From Table I it is evident that for all plasmid constructs and at all Hg^{2+} concentrations, the transport rate $r_{T,max}$ is several times slower than the reaction rate $r_{R,max}$. Similarly, r_T is significantly slower than r_R (30). This is an indication that the Hg^{2+} transport is the rate-determining step of the Hg^{2+} reduction process at the whole cell level. For all gene copy numbers, both the transport and the reaction rate exhibit their maximum at an extracellular Hg^{2+} concentration of about 40 μM; at that concentration, r_R is 3.5-fold faster than r_T for the R100-harboring cells and 5.5- to almost 10-fold faster than r_T for the higher copy number plasmids (32).

In conclusion, the developed mathematical model can satisfactorily predict r_T and r_R for various copy numbers of the *mer* operon in a wide range of substrate (Hg^{2+}) concentrations. This emphasizes the applicability of the model and justifies its use as a means for the design of a better biocatalyst.

STRATEGIES FOR OPTIMIZATION OF MERCURIC REDUCTION
In order to correlate $r_{T,max}$ and $r_{R,max}$ with the respective polypeptide concentrations specified by the different operon copy numbers, the relative concentrations of the transport proteins and mercuric reductase were determined using a combination of maxicells, fluorography, and one-dimension video densitometry (32).

The effect of gene amplification on the Hg^{2+} transport rate and the concentration of the transport proteins is shown in Figure 6. Both the rate and the concentration initially increased with gene copy number, but were saturated at higher copy numbers following similar patterns. The increase of the transport rate was proportional to that of the transport protein concentration throughout the gene copy number amplification range. Overall, the 47-fold gene amplification resulted in only a 2.5-fold increase of the transport rate. Figure 6 suggests that the lower than expected increase of the transport rate, and therefore of the Hg^{2+} reduction rate by intact cells, is due to the modest amplification of the transport protein levels in the cell. In contrast, both the reaction rate and the concentration of mercuric reductase increased linearly with gene copy number (Figure 7). The 47-fold gene amplification resulted in an overall 7-fold increase in the reaction rate, which correlates with the 5-fold amplification of the intracellular enzyme concentration.

Amplification of the entire *mer* operon did not considerably increase the reduction activity of the intact cells. Amplification of the transport rate appears to be the limiting factor in that optimization approach. Presumably both the transport and the reductase genes on the same plasmid are amplified to the same extent, but the protein levels of these two genes are different. This difference could be caused by different gene transcription or translation level or by differential turnover rates of the mRNAs within the *mer* operon. Conceivably, separation (by subcloning) of the transport genes from the reductase gene such that expression is controlled by different transcription regulatory systems may provide the means for further amplification of the transport protein levels. Thus, an alternative strategy towards optimization of the Hg^{2+} reduction rate can be developed based on modification of the ratio of transport protein concentration to mercuric reductase concentration. The goal of such a manipulation will be to increase the intracellular Hg^{2+} concentration to an optimal, but not inhibitory, level and allow the cytoplasmic enzyme to function at a faster turnover rate (Equation 4).

Based on the assumption of steady state between the transport rate and the reaction rate at the whole-cell level, Equations 4 and 6 can be equated:

$$r_R = r_T \quad \text{or} \quad \frac{r_{R,max}\,[Hg^{2+}]_i}{K_m + [Hg^{2+}]_i + [Hg^{2+}]_i^2/K_i} = \frac{r_{T,max}\,[Hg^{2+}]_o}{K_1 + [Hg^{2+}]_o} \tag{7}$$

Equation 7 can be solved in terms of the intracellular Hg^{2+} concentration, $[Hg^{2+}]_i$:

$$[Hg^{2+}]_i = \frac{F - (F^2 - 4K_m K_i)^{1/2}}{2} \tag{8}$$

where $F = ((K_1 + [Hg^{2+}]_o)/[Hg^{2+}]_o/(r_{T,max}/r_{R,max})-1)K_i$. The other root of the equation is rejected, since it yields $[Hg^{2+}]_i$ greater than $[Hg^{2+}]_o$. Using Equation 8, the intracellular Hg^{2+} concentrations in cells carrying the five *mer* plasmids were determined at 40 μM of extracellular Hg^{2+} concentration (Table II).

Despite the variation of $r_{T,max}$ and $r_{R,max}$, as a result of gene amplification, the corresponding intracellular Hg^{2+} concentrations fall in a relatively narrow range below 5 μM, which may reflect the maximal intracellular Hg^{2+} concentration that cells can tolerate (Table II). By maintaining $[Hg^{2+}]_i$ at a low level, cells utilize only a fraction

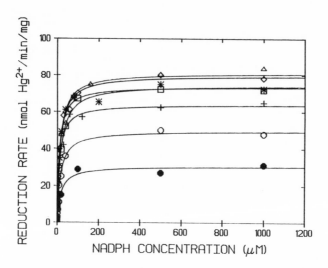

Figure 5. Dependence of the Hg^{2+} reduction rate on the NADPH concentration in the presence of various concentrations of Hg^{2+}. Concentrations of Hg^{2+}: (●): 5 µM; (o): 10 µM; (+): 20 µM; (Δ): 40 µM; (◇): 60 µM; (*): 80 µM; (□): 120 µM. The solid lines represent the predictions of Equation 4 (from Ref. 29).
(Reproduced with permission from reference 29. Copyright 1990 Butterworth–Heinemann.)

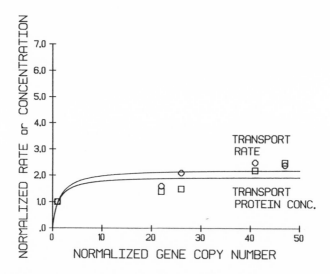

Figure 6. Effect of gene amplification on the transport protein concentration and the Hg^{2+} transport rate. The normalized Hg^{2+} transport rates (□) were calculated from the data in Table I. The normalized transport protein concentration (O) were derived from a fluorogram of labeled polypeptides produced in maxicells harboring the various copy number plasmids.

Table II. Intracellular Hg^{2+} concentration ($[Hg^{2+}]_i$) and
values of the $r_T/r_{T,max}$ and $r_R/r_{R,max}$ ratios for *E. coli* cells
harboring the *mer* plasmids, in the presence of 40 μM extracellular Hg^{2+} [a]

PLASMID	$r_T/r_{T,max}$	$r_R/r_{R,max}$	$[Hg^{2+}]_i$ (μM)
R100	0.91	0.17	3.43
pBRmer	0.89	0.08	1.48
pDU1003	0.87	0.09	1.78
pACYCmer	1.00	0.09	1.57
pUCmer	0.89	0.06	1.07

[a] Compiled from data in Ref. 32.

of their turnover capacity of mercuric reductase (from 6 to 17% of $r_{R,max}$; Table II), although they operate close to their maximal transport capacity (Table II). Thus, rapid detoxification of the intracellular Hg^{2+} is ensured, while the cells reserve the ability to cope with an increased Hg^{2+} concentration that could result from a sudden surge in extracellular Hg^{2+} concentration. An increased extracellular Hg^{2+} concentration would lead to an increased rate of Hg^{2+} uptake. With the large capacity of mercuric reductase in reserve, this increased Hg^{2+} transport rate may not result in a sudden increase of intracellular Hg^{2+} concentration. Thus, in its natural habitat, the cell seems to have derived a mechanism for maintaining low intracellular Hg^{2+} concentration over a wide range of extracellular Hg^{2+} concentration. Such a mechanism provides an advantage to the survival of the cell. However, from a biocatalytic standpoint, that provision limits the productivity of the Hg^{2+} reduction process.

Equation 8 expresses the intracellular Hg^{2+} concentration as a function of the extracellular Hg^{2+} concentration and the term $r_{T,max}/r_{R,max}$. That term represents the ratio of the transport protein concentration to the concentration of mercuric reductase. Equation 8 predicts an increase in $[Hg^{2+}]_i$ with increasing $r_{T,max}/r_{R,max}$. The ratio $r_{T,max}/r_{R,max}$ can, in principle, be manipulated by amplifying the expression of the transport and the reductase genes to different extents. The predicted effect of changing this ratio on $[Hg^{2+}]_i$ is depicted in Figure 8 for various extracellular Hg^{2+} concentrations, $[Hg^{2+}]_o$ (27). The $[Hg^{2+}]_i$ values calculated for $[Hg^{2+}]_o = 40$ μM for the five plasmid constructs are also shown in Figure 8. The values of K_1, K_m, and K_i used in Equation 8 were the mean ones for the five recombinant plasmid-harboring cell constructs (4.6, 17.3, and 88.8 μM, respectively). At low $[Hg^{2+}]_o$ the intracellular Hg^{2+} concentration increases linearly, but slowly. However, at $[Hg^{2+}]_o$ beyond 10 μM, the intracellular Hg^{2+} concentration increases rapidly with the increasing $r_{T,max}/r_{R,max}$ ratio.

The presented alternative approach to optimization of the Hg^{2+} reduction rate at the whole-cell level is subject to certain limitations. The first limitation is a critical value of the ratio $r_{T,max}/r_{R,max}$, designated $(r_{T,max}/r_{R,max})_{cr}$, beyond which no solution to the steady state Equation 7 is feasible. The critical ratio decreases with increasing extracellular Hg^{2+} concentration, $[Hg^{2+}]_o$. For $[Hg^{2+}]_o$ values between 10 and 120 μM it ranges from 0.8 to about 0.6.

As can be seen from Figure 8, increasing $r_{T,max}$ preferentially over $r_{R,max}$ achieves a higher intracellular Hg^{2+} concentration and the reductase is allowed to operate at a higher turnover rate. So, to optimize Hg^{2+} reduction by recombinant *E. coli* cells, the transport proteins and mercuric reductase need to be amplified to different extents in

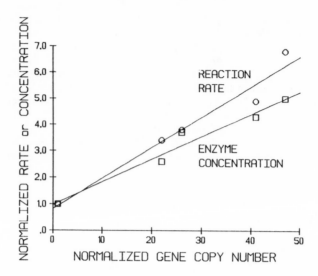

Figure 7. Effect of gene amplification on the mercuric reductase concentration and the Hg^{2+} reaction rate. The normalized Hg^{2+} reaction rates (□) were calculated from the data in Table I. The normalized concentrations of mercuric reductase (O) were derived from maxicell expression of the reductase on the different copy numbered plasmids.

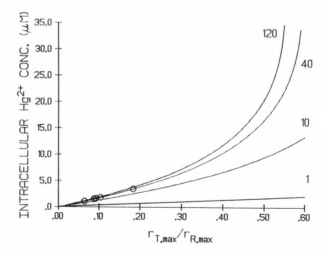

Figure 8. Effect of the $r_{T,max}/r_{R,max}$ ratio on the intracellular Hg^{2+} concentration in the presence of various extracellular Hg^{2+} concentrations. Each curve has been labeled by the corresponding extracellular Hg^{2+} concentration (in μM). The data points correspond to cells carrying the five recombinant *mer* plasmids in 40 μM extracellular Hg^{2+} concentration (Table II) (from Ref. 27). (Reproduced with permission from reference 27. Copyright 1991 John Wiley & Sons.)

order to achieve a favorable $r_{T,max}/r_{R,max}$ ratio. However, the optimal value of $[Hg^{2+}]_i$ is limited by the tolerance of the cell to high intracellular Hg^{2+} concentrations. At relatively high values of the $r_{T,max}/r_{R,max}$ ratio, the intracellular Hg^{2+} concentration increases to significant levels with increasing extracellular Hg^{2+} concentration (Figure 8). Therefore, although high $[Hg^{2+}]_i$ will result in increased reduction rates, the $[Hg^{2+}]_i$ should not exceed a certain level, $[Hg^{2+}]_{i,cr}$, beyond which the viability of the cell is affected. The value of $[Hg^{2+}]_{i,cr}$ is a second limitation on optimization of the $r_{T,max}/r_{R,max}$ ratio. Determination of that critical level of $[Hg^{2+}]_i$ will require an investigation of the effect of Hg^{2+} concentration on cell viability.

When superimposed on the predictions of the model, the two limitations can help determine the optimal ratio of the concentrations of transport proteins and mercuric reductase, $(r_{T,max}/r_{R,max})_{opt}$, which ensures performance of the cell at an optimal reduction rate. From the position of the data corresponding to the five *mer* plasmids in Figure 8, it seems that there is a wide margin of possible improvement in the biocatalytic performance of recombinant cells.

ACKNOWLEDGEMENTS
This work was supported in part by grants from Ecolab, Inc. (St. Paul, MN), the National Science Foundation (ECE-8552670), and the Graduate School of the University of Minnesota. The authors wish to thank Professor Simon Silver, University of Illinois, for valuable suggestions.

NOMENCLATURE

A	Cell envelope area available for Hg^{2+} transport
D_j, D'_j	Effective diffusion coefficient (permeability)
E	Mercuric reductase (*merA* gene product)
F	Term defined in Equation 8
k_j	Reaction rate constant ($j = 1$-12), introduced in Figure 3
K_i', K_i'', K_{sA}, K_{mA}, K_{mB}	Parameters of the reduction rate expression, defined in Equation 3
K_1	Dissociation constant of the P-Hg complex
K_2	Dissociation constant of the T-Hg complex
K_m, K_i	Parameters of the reduction rate Equation 4
P	Periplasmic protein (*merP* gene product)
r_o	Overall Hg^{2+} reduction rate by whole cell
r_R	Initial Hg^{2+} reduction rate
$r_{R,max}$	Maximal rate of reduction, defined in Equation 4
$r'_{R,max}$	Maximal rate of reduction, defined in Equation 3
r_T	Initial Hg^{2+} transport rate across the cellular envelope
$r_{T,max}$	Maximal Hg^{2+} transport rate across the cellular envelope, defined in Equation 2
S-Hg	Species S-mercuric ion complex
S-NADP(H)	Species S-NADP(H) complex
S-NADP(H)-Hg	Species S complexed by NADP(H) and mercuric ion
[S]	Concentration of species S
$[S]_t$	Total concentration of species S
T	Inner membrane protein (*merT* gene product)
V	Volume of the cell

Subscripts

i	Intracellular
o	Extracellular

LITERATURE CITED

1. Summers, A.O.; Silver, S. Annu. Rev. Microbiol. 1978, 32, 637-672.
2. Ni'Bhriain, N.N.; Silver, S.; Foster, T.J. J. Bacteriol. 1983, 155, 690-703.
3. Jackson, W.J.; Summers, A.O. J. Bacteriol. 1982, 151, 962-970.
4. Ni'Bhriain, N.; Foster, T.J. Gene 1986, 42, 323-330.
5. Foster, T.J.; Nakahara, H.; Weiss, A.A.; Silver, S. J. Bacteriol. 1979, 140, 167-181.
6. Barrineau, P.; Gilbert, P.; Jackson, W.J.; Jones, C.S.; Summers, A.O.; Wisdom, S. J. Mol. Appl. Genet. 1984, 2, 601-619.
7. Misra, T.K.; Brown, N.L.; Fritzinger, D.C.; Pridmore, R.D.; Barnes, W.M.; Haberstroh, L.; Silver, S. Proc. Natl. Acad. Sci. USA 1984, 81, 5975-5979.
8. Foster, T.J.; Brown, N.L. J. Bacteriol. 1985, 163, 1153-1157.
9. O'Halloran, T.; Walsh, C. Science 1987, 235, 211-214.
10 Brown, N.L., Misra, T.K., Winnie, J.N., Schmidt, A., Seiff, M. and Silver, S. Mol. Gen. Genet. 1986, 202, 143-151.
11. O'Halloran, T.; Frantz, B.; Shin, M.K.; Ralston, D.M.; Wright, J.G. Cell 1989, 56, 119-129.
12. Summers, A.O. Annu. Rev. Microbiol., 1986, 40, 607-634.
13. Lund, P.A.; Brown, N.L. Gene 1987, 52, 207-214.
14. Misra, T.K.; Brown, N.L.; Haberstroh, L.; Schmidt, A.; Goddette, D.; Silver S. Gene 1985, 34, 253-262.
15. Summers, A.O.; Sugarman, L.I. J. Bacteriol. 1974, 119, 242-249.
16. Nakahara, H.S.; Silver, S.; Miki, T.; Rownd, R.H. J. Bacteriol. 1979, 140, 161-166.
17. Schottel, J.L. J. Biol. Chem. 1978, 253, 4341-4349.
18. Fox, B.; Walsh, C.T. J. Biol. Chem. 1982, 257, 2498-2503.
19. Schultz, P.G.; Au, K.G.; Walsh, C.T. Biochemistry 1985, 24, 6840-6848.
20. Brown, N.L.; Ford, S.J.; Pridmore, R.D.; Fritzinger, D.C. Biochemistry 1983, 22, 4089-4095.
21. Distefano, M.D.; Au, K.G.; Walsh, C.T. Biochemistry 1989, 28, 1168-1183.
22. Moore, M.J.; Walsh, C.T. Biochemistry 1989, 28, 1183-1194.
23. Miller, S.M.; Moore, M.J.; Massey, V.; Williams, C.H. Jr.; Distefano, M.D.; Ballou, D.P.; Walsh, C.T. Biochemistry 1989, 28, 1194-1205.
24. Miller, S.M.; Ballou, D.P.; Massey, V.; Williams, C.H. Jr.; Walsh, C.T. J. Biol. Chem. 1986, 261, 8081-8084.
25. Rinderle, S.J.; Booth, J.E.; Williams, J.W. Biochemistry 1983, 22, 869-876.
26. Summers, A.O. and Sugarman, L.I. J. Bacteriol. 1974, 119, 242-249.
27. Philippidis, G.P.; Schottel, J.L.; Hu, W.-S. Biotechnol. Bioeng. 1991, 37, 47-54.
28. Dixon, M.; Webb, E.C. Enzymes; Academic Press: New York, NY, 1979; pp. 126-136.
29. Philippidis, G.P.; Schottel, J.L.; Hu, W.-S. Enzyme Microb. Technol. 1990, 12, 854-859.
30. Bochner, B.R.; Ames, B.N. J. Biol. Chem. 1982, 257, 9759-9769.
31. Ames, G.F.-L. Ann. Rev. Microbiol. 1986, 55, 397-425.
32. Philippidis, G.P. Ph.D. Thesis, University of Minnesota, Minneapolis, 1989.

RECEIVED June 26, 1991

Chapter 5

Ubiquitin Fusion Approach to Heterologous Gene Expression in Yeast

High-Level Production of Amino-Terminal Authentic Proteins

Philip J. Barr, Elizabeth A. Sabin[1], Chun Ting Lee Ng,
Katherine E. Landsberg, Kathelyn S. Steimer, Ian C. Bathurst,
and Jeffrey R. Shuster

Chiron Corporation, 4560 Horton Street, Emeryville, CA 94608

The ubiquitin fusion approach to gene expression in eukaryotic systems allows the production of heterologous proteins that are cleaved precisely *in vivo* from the ubiquitin fusion partner by an endogenous ubiquitin-specific hydrolase. Alternatively, ubiquitin fusions can be isolated from bacterial hosts, and cleaved by the hydrolase *in vitro*. In each case, the system gives recombinant proteins that contain authentic amino termini. We have found the system to be of particular utility for the production of human gamma interferon (γIFN) and α_1-proteinase inhibitor (α_1-PI). We have also used the system for the production of regions of the human immunodeficiency virus type-1 (HIV1) genome that were previously expressed at only low levels in yeast. These include domains of the HIV1 *env* gene and also the region of the HIV1 *pol* gene that encodes the HIV1 integrase enzyme. Surprisingly, for one of the *env* proteins we were able to isolate a product in which the amino-terminal Glu residue was modified by addition of an Arg residue. This arginyl-tRNA-protein-transferase catalyzed process had previously only been observed for short-lived intermediates in the ubiquitin-dependent proteolytic degradation pathway. The HIV1 integrase was found to contain the authentic amino terminus (Phe.Leu.Asn...) previously reported for virion-associated integrase.

The eukaryotic protein ubiquitin currently represents an area of intense study. The 76 amino acid ubiquitin polypeptide is highly conserved in all eukaryotes studied, and has been shown to be involved in many important areas of cell biology. In mammalian systems, ubiquitin has been located at the cell surface, covalently bound to the membrane associated platelet-derived growth factor receptor *(1)* and the lymphocyte homing receptor *(2)*. Covalently bound or free ubiquitin is also found in the cytoplasm and cell nucleus *(3,4)*.

Genetic approaches to the study of ubiquitin function in yeast have led to major advances in our understanding of the role of ubiquitin in the selective or

---

[1]Current address: Department of Veterinary Microbiology and Immunology, University of California—Davis, Davis, CA 95616

0097–6156/91/0477–0051$06.00/0

targeted degradation of intracellular proteins *(5-8)*. Furthermore, this molecular approach has led to the discovery of novel functions for ubiquitin fusion proteins encoded by each discrete ubiquitin gene of *S. cerevisiae (9)*. An important discovery in the ubiquitin system has been the identification of a requirement for proteolytic maturation of each ubiquitin gene product. Polyubiquitin, the product of the *UBI4* gene in yeast, is a pentameric precursor *(10)*, whereas *UBI1*, *UBI2* and *UBI3* encode ubiquitin-hybrid proteins with unrelated 'tail' sequences *(11)*. The *UBI4* gene is heat stress inducible and has been shown to be an essential component of heat-shock response mechanisms *(6)*. The other members of the multigene family have been found recently to encode ribosomal proteins whose fusion to ubiquitin facilitates ribosome biogenesis *(9)*. In each case, proteolysis of the hybrid proteins has been shown to be extremely rapid, efficient, and specific for cleavage of the peptide bond subsequent to Gly76 of ubiquitin. A yeast hydrolase responsible for this cleavage has been characterized at the molecular level by cloning and overexpression of its gene product *(12)*. These new findings have been reviewed recently with particular regard to the structure and function of ubiquitin *(13)*.

Subsequent to the discovery that the ubiquitin-specific hydrolase could also cleave heterologous protein hybrids *in vivo (5)* the system has been utilized for the production of recombinant proteins with specifically, engineered amino termini. Initially, production of bacterial β-galactosidase derivatives and murine dihydrofolate reductases (DHFRs) that differed exclusively at their amino-terminal residues led to the definition of the N-end rule *(5,7,8)*. According to this general rule, specific amino acids can be ranked according to the degree of stabilization, or destabilization, that they confer upon proteins when positioned at their amino termini. More recently, we *(14, 15)*, and others *(16,17)*, have extended these observations and, using strong and regulatable promoters, have produced high levels of heterologous eukaryotic proteins in yeast. A significant advantage to this approach is that for potential therapeutic products, and for other proteins where amino-terminal authenticity is desirable or critical, the obligatory methionine initiation codon is situated at the amino terminus of ubiquitin, and not attached to the heterologous protein. In contrast to direct intracellular expression, therefore, where this potentially immunogenic methionine residue is frequently encountered, the processed heterologous protein can be engineered for amino-terminal authenticity *(15)*. A second, and perhaps less expected finding has been that, as with traditional fusion protein expression strategies in which a highly stable fusion partner confers stability on otherwise unstable polypeptides, the ubiquitin fusion approach has also led to greatly increased levels of expression of quantitatively processed heterologous proteins *(14,17)*.

S. cerevisiae has been demonstrated to be an attractive host system for the expression of genes of the human immunodeficiency virus (HIV). We have shown previously, that regions of the *gag, pol* and *env* genes of HIV1 can be expressed at high levels in yeast. The *gag* polyprotein precursor p53, synthesized in yeast, is amino-terminally myristylated, as is its natural counterpart isolated from virions *(18)*. Active recombinant enzymes from the HIV1 *pol* gene include the active aspartyl proteinase domain, secreted from yeast using the α-factor mating pheromone leader sequence *(19)*, and reverse transcriptase (RT) *(20)*. Also, a human superoxide dismutase (hSOD) fusion protein containing regions of the HIV1 integrase, an enzyme encoded by the 3'-end of the *pol* gene, has found utility, along with *gag* and *env* gene products, in the detection of antibodies to the AIDS virus in infected individuals *(21)*. Lastly, several domains of the *env* gene have been expressed in yeast either directly or as hSOD fusion proteins *(23)*. A non-glycosylated equivalent of the exterior domain of the HIV1 envelope, gp120, has been shown to induce a variety of immune responses in experimental animals, and is currently in human clinical trials as a vaccine against AIDS *(22,23)*.

It is clearly not optimal for a potential immunogen to be expressed as a fusion protein. Similarly, for an enzyme, such as the HIV1 integrase, where active enzyme might be useful for structural and activity studies aimed towards the chemotherapy of AIDS, the most authentic recombinant product possible is desirable. In this paper, therefore, we extend the use of the ubiquitin fusion approach, and focus primarily on regions of the HIV1 genome that have previously been difficult to express in yeast.

Experimental Procedures

Nucleic Acid Manipulations. Oligonucleotides were synthesized by the phosphoramidite method using Applied Biosystems 380A DNA synthesizers. Construction of the synthetic ubiquitin gene, the parent yeast expression vector, and the human γ-interferon (hγ-IFN) and α_1-proteinase inhibitor (α_1PI) fusion genes has been described previously *(15)*. HIV1 DNA was obtained from previously manipulated subclones of the proviral genome of the SF2 strain of HIV1, into which Sal-1 sites had been inserted immediately subsequent to the relevant termination codons *(22,24)*. The gene for env4 encodes amino acids 272 to 509 of the *env* gene product (25), and therefore, corresponds to the carboxyl-terminal half of gp120. The longer env4-5 gene encodes amino acids 272 to 673 of the *env* gene product and, therefore, includes 174 amino acids of gp41. The ubiquitin junction sequences were made using a unique Bgl-2 site for env4 and env4-5 together with the synthetic adapters shown in Figure 1 (e,f). Similarly, a synthetic adapter was used to fuse the ubiquitin and integrase genes at the Sst-II site of ubiquitin and an engineered Cla-1 site at Ile5 of integrase (Figure 1(g)) *(25,26)*. All synthetic DNA constructions were verified by M13 dideoxy sequencing (26), and plasmid preparations and enzyme reactions were essentially as described *(27)*. Restriction enzymes, T4 polynucleotide kinase and T4 DNA ligase were obtained from NEB and BRL. For northern blot analysis, yeast RNA was isolated from saturated yeast cultures, electrophoretically separated (20µg per lane), and blotted as described *(28,29)*. Hybridization to a 5'-^{32}P labeled oligonucleotide probe (40mer) complementary to the 5'-untranslated sequence of the glyceraldehyde-3-phosphate dehydrogenase (GAPDH) mRNA allowed for quantitation of RNA levels using the chromosomal GAPDH gene messages as an internal control. Approximate relative message levels were obtained by densitometric scanning using a Shimadzu CS9000 scanning densitometer.

Bacterial and Yeast Strains. *E. coli* strains HB101 and D1210 were used for gene cloning and plasmid amplification. For yeast expression *S. cerevisiae* strain AB116: *(MATa leu2 trp1 ura3-52 prB1-1122 pep4-3 prC1-407* [cir°]) and JSC302, an *ADR1* overexpressing derivative of AB116 *(30)* were used exclusively. Yeast cells were transformed by the spheroplast method *(31)* and propagated under conditions of leucine selection prior to liquid culture in YEP media *(32)*. Induction of heterologous gene expression was as described previously *(32)*.

Analysis of Expressed Protein Products. Induced yeast cultures were harvested by centrifugation, and cells analyzed by SDS-PAGE on 12.5% or 15-22% gradient gels with Coomassie blue staining as described *(15)*. For envelope and integrase proteins, lysis was by the glass bead method *(24)* in Triton lysis

```
(a)   ARG.GLY.GLY.STOP.                                        pBS24Ub
      C.CGC.GGT.GGT.TAG.TCG.AC
      G.GCG.CCA.CCA.ATC.AGC.TG
        Sst-II            Sal-1


              ↓
(b)   ARG.GLY.GLY.GLN.ASP.PRO.TYR.VAL.LYS.GLU...              pBS24Ubhγ-IFN
      C.CGC GGT.GGT.CAG GAT.CCA.TAC.GTT.AAG.GAA
      G.GCG.CCA.CCA.GTC.CTA.GGT.ATG.CAA.TTC.CTT
                         BamH1


              ↓
(c)   ARG.GLY.GLY.ARG.ASP.PRO.TYR.VAL.LYS.GLU...              pBS24Ubhγ-IFNQ1R
      C.CGC GGT.GGT.AGA GAT.CCA.TAC.GTT.AAG.GAA
      G.GCG.CCA.CCA.TCT.CTA.GGT.ATG.CAA.TTC.CTT
                      BamH1 overhang


              ↓
(d)   ARG.GLY.GLY.GLU.ASP.PRO.GLN.GLY.ASP.ALA...              pBS24Ubα1-PI
      C.CGC GGT.GGC.GAA GAT.CCC.CAG.GGA.GAT.GCT
      G.GCG.CCA.CCG.CTT.CTA.GGG.GTC.CCT.CTA.CGA
                      BamH1 overhang


              ↓
(e)   ARG.GLY.GLY.GLU.VAL.VAL.ILE.ARG.SER.ASP...              pBS24Ubenv4
      C.CGC GGT.GGC.GAG.GTA.GTA.ATT.AGA.TCT.GA
      G.GCG.CCA.CCG.CTC.CAT.CAT.TAA.TCT.AGA.CT
                                  Bgl-2


              ↓
(f)   ARG.GLY.GLY.MET.GLU.VAL.VAL.ILE.ARG.SER...              pBS24Ubenv4-5
      C.CGC GGT.GGC.ATG.GAG.GTA.GTA.ATT.AGA.TCT
      G.GCG.CCA.CCG.TAC.CTC.CAT.CAT.TAA.TCT.AGA
                                  Bgl-2


              ↓
(g)   ARG.GLY.GLY.PHE.LEU.ASN.GLY.ILE.ASP.LYS...              pBS24UbIntegrase
      C.CGC GGT.GGC.TTT.TTG.AAT.GGT.ATC.GAT.AAG
      G.GCG.CCA.CCG.AAA.AAC.TTA.CCA.TAG.CTA.TTC
                                  Cla-1
```

FIG. 1.   Synthetic DNA (boxed) and encoded amino acid sequences at the ubiquitin-heterologous gene and protein junctions. (a) The parent vector pBS24Ub contains a unique Sst-II site close to the 3'-end of the ubiquitin gene, and a unique Sal-1 site immediately subsequent to the termination codon. These restriction sites, together with the cloning sites shown for each heterologous gene, are used for construction of subsequent Ub-fusion vectors (b-g). The ubiquitin hydrolase cleavage site for each fusion protein is shown (arrowed).

buffer (0.1M sodium phosphate pH7.3, 1.25mM EDTA, 0.1% Triton X-100). For gel analysis, the insoluble fraction was boiled for 5 min. in sample loading buffer (62.5 mM Tris-Cl pH 6.8, 50 mM DTT, 3% SDS, 10% glycerol) prior to loading. Amino-terminal amino acid sequence analysis was performed by the Immobilon-P membrane method *(32)* as described previously *(15)*, using an Applied Biosystems 470A gas phase sequencer.

Quantitation of Expression of HIV1 Proteins. Comparative assessment of expression levels from direct, ubiquitin fusion, or hSOD fusion expression systems was obtained using a quantitative slot blot method. Reagents used for detection were from the commercially available RIBA AIDS strip kit (Chiron Corporation/Ortho Diagnostics). 1 ml samples of induced yeast cells were centrifuged and lysed by boiling for 5 min in 200 μL of SDS lysis buffer (150 mM Tris-Cl pH 6.8, 2% SDS, 3% 2-mercaptoethanol). The samples were centrifuged, and the supernatants diluted (1:62) in strip kit coating solution *(21)* prior to loading onto nitrocellulose (Schleicher and Schuell) using a slot blot apparatus (BioRad). The equivalent of 10, 2, 0.4, 0.08, and 0.016 mL of culture were loaded and for comparison, previously quantitated standards of purified recombinant HIV1 envelopes *(22)* and SODp31 *(24)*, an hSOD fusion protein that contains a large portion of the HIV1 integrase. The nitrocellulose was air-dried, washed overnight with RIBA kit wash buffer, and then probed for 30 min with pooled HIV positive sera from infected humans (1:200 dilution). Blots were developed, and scanned using the reflective mode on a Shimadzu CS 9000 scanning densitometer. Peaks were integrated, compared with the appropriate standards, and approximate expression levels determined (Table I).

TABLE I. Expression levels of HIV1 proteins produced in yeast. Cultures were harvested after 24h growth in YEP media containing 2% glucose and analyzed as described in "Experimental Procedures". Background signals from yeast cells transformed with a control plasmid were undetectable

Recombinant protein	Expression system	Expression level (mg/l)
env4	direct	17
	hSOD fusion	35
	Ub fusion	75
env4-5	direct	6
	hSOD fusion	7
	Ub fusion	22
HIV1 integrase	direct	28
	Ub fusion	122

Results

Expression Vectors.   Schematic representations of the parent ubiquitin expression vector (pBS24Ub), and the derived fusion vectors are shown (Figure 2).   In all cases, transcription is driven by the alcohol dehydrogenase 2/glyceradehyde-3-phosphate dehydrogenase (ADH2/GAPDH) promoter described previously (15,30,32).   Also shown (Figure 1), are the synthetic DNA junction sequences and their encoded protein sequences for the various ubiquitin fusion constructions.   Corresponding direct expression constructions were engineered to commence at identical amino acid positions, after translation initiation on a methionine codon.

Heterologous Protein Expression.   Yeast cells transformed with the above plasmids were induced for expression by growth and concomitant depletion of glucose from the culture media.   Expression results, using Coomassie blue stained polyacrylamide gels, are shown in Figure 3.   The parent plasmid, pBS24Ub is clearly capable of producing high levels of yeast ubiquitin (Figure 3(a), lane 2).   Similarly, when fused to ubiquitin, high levels of active hγ-IFN and α_1PI are produced by quantitative *in vivo* cleavage of each fusion protein (Figure 3(a)), lanes 3 and 5).   Direct expression of these two proteins also gives high levels of each active protein (Figure 3(a), lanes 4 and 6).   As we have described previously however (15), the directly expressed proteins contain the undesirable, initiation codon-derived methionine residues at their amino termini.   When the amino terminal glutamine residue of hγ-IFN was replaced with arginine (Figure 1 (c)), a highly destabilizing residue according to the N-end rule, we observed equally high levels of the Triton lysis buffer insoluble hγ-IFNQ1R analog by SDS-PAGE analysis.

Coomassie blue stained SDS-PAGE results for the HIV1 envelope polypeptide env4, the carboxyl-terminal half of gp120, and the HIV1 integrase are shown in Figure 3(b).   For env4, direct expression (lane 2) is low , whereas both ubiquitin fusion (lane 3) and hSOD fusion expression (lane 4) is considerably higher (see also Table I).   Similarly, for the longer envelope polypeptide env4-5, that contains a large portion of gp41 sequences, expression levels were significantly enhanced when the env4-5 gene was fused to either the ubiquitin or hSOD genes (Table I).   For HIV1 integrase, we also compared direct expression with ubiquitin fusion expression.   As evident from SDS-PAGE (Figure 3(b); lanes 5 and 6) and quantitative immunoblot analysis (Table I), the use of ubiquitin as an *in vivo* cleaved fusion partner for HIV1 integrase clearly gives superior levels of viral protein production.

Protein Purification and Amino-Terminal Analysis.   Specific cellular mechanisms exist for the amino terminal modification of proteins prior to degradation *via* the N-end rule pathway.   It has been shown previously (6,8,9), that amino-terminal Gln and Asn residues can be hydrolyzed to Glu and Asp residues respectively.   Gln and Asn are referred to as tertiary destabilizing residues (6,8,9).   The secondary destabilizing residues, Glu, Asp and Cys can be modified by arginyl-RNA-protein-transferase catalyzed addition of Arg, a primary destabilizing amino acid to the amino terminus of a substrate protein.   The addition of Arg allows binding of the target protein by the ubiquitin-conjugating enzyme (8,9), ubiquitination of internal lysine residues and subsequent rapid degradation.   The existence of these modification pathways has several implications for heterologous gene expression using the ubiquitin fusion approach.   We therefore determined the amino-terminal amino acid sequences of several of the expressed proteins (Table II).   Although hγ-IFN and α_1-PI have tertiary (Gln) and secondary (Glu) destabilizing amino acids respectively at

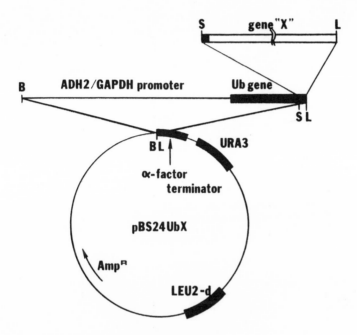

FIG. 2.   Schematic representation of the ubiquitin fusion expression plasmid. The parent vector pBS24Ub contains selectable markers for yeast growth in uracil- or leucine- deficient media. The promoter-ubiquitin "cassette" is cloned as a BamH1(B), Sal-1(L) fragment. Heterologous genes (X) are cloned into pBS24Ub as Sst-II(S)/Sal-1 fragments. Transcription is driven by the ADH2/GAPDH promoter described previously *(15,30,31)*. Precise junction sequences are shown in Figure 1.

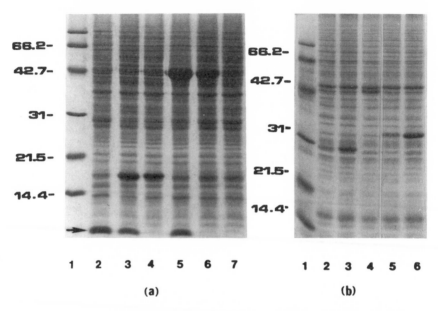

FIG. 3.   Coomassie blue stained 15-22% gradient (a) and 12.5% (b)
SDS-polyacrylamide gels of heterologous proteins expressed in yeast.
Transformed yeast cells were cultured, lysed and analyzed as described in
"Experimental Procedures". (a) Lane 1, molecular weight markers
(BioRad); Lane 2, yeast ubiquitin (arrowed); Lane 3, hγ-IFN and ubiquitin
derived from *in vivo* cleavage of the fusion protein; Lane 4, N-methionyl-
hγ-IFN; Lane 5, α_1-PI and ubiquitin from *in vivo* cleavage of the fusion
protein; Lane 6, N-methionyl-α_1-PI; Lane 7, control lysate from yeast
cells transformed with the parent plasmid pBS24. (b) Lane 1, molecular
weight markers; Lane 2, directly expressed env4; Lane 3, ubiquitin
fusion-derived env4, clearly visible close to its calculated molecular
weight of 27.1 kD.   Lane 4, hSOD-env4 fusion, molecular weight 42.9 kD;
Lane 5, directly expressed HIV1 integrase, and Lane 6, ubiquitin fusion-
derived HIV1 integrase of calculated molecular weight 32.3 kD.

their amino termini, we have shown previously that the expressed products contain exclusively the amino acids predicted from the DNA sequence in this position. In contrast, however, sequence analysis of env4 revealed that only approximately 30% of the recombinant envelope polypeptide initiated with Glu. The remainder contained an additional Arg residue preceding the Glu residue predicted from the DNA sequence of the cleaved protein (Table II).

TABLE II. Amino terminal amino acid sequence of recombinant proteins predicted from DNA sequences and the known specificity of the ubiquitin-specific hydrolase *(10,12)*, and experimentally determined sequences of purified proteins

Protein	Predicted Amino-Terminal Amino Acid Sequence	Experimentally Determined Amino-Terminal Amino Acid Sequence	
	1	1	
hγ-IFN	Gln.Asp.Pro....	Gln.Asp.Pro....	100%
α_1-PI	Glu.Asp.Pro....	Glu.Asp.Pro....	100%
env4	Glu.Val.Val....	Arg.Glu.Val....	70%
		Glu.Val....	30%
Integrase	Phe.Leu.Asn....	Phe.Leu.Asn....	100%

The amino terminus of virion-associated HIV1 integrase has been shown to be Phe716 of the *pol* open reading frame *(25,26)*. Phe belongs to the subset of amino acids that, when situated next to the initiation codon-derived Met residue, inhibit its removal by methionine aminopeptidase *(34)*. Thus, it is likely that the directly expressed HIV1 integrase retains this additional Met residue. Sequence analysis of ubiquitin fusion-derived integrase, gave an amino terminus of Phe exclusively (Table II).

mRNA Levels of Heterologous Transcripts. In order to define possible mechanisms for the enhanced levels of expressed proteins using the ubiquitin fusion approach, we measured mRNA levels for directly expressed env4 and integrase, the corresponding ubiquitin fusions, and the hSOD-env4 fusion. Results are shown in Figure 4. For integrase, a correlation exists between message levels and the enhanced expression from the ubiquitin fusion construction. The ubiquitin fusion message levels were approximately 4.5 fold higher than for direct expression, comparing favorably with the 4.4 fold enhancement of protein levels observed. In contrast, however, no correlation existed with env4 mRNA levels. Indeed, message levels from the direct env4 construction are even marginally higher than from the ubiquitin fusion construction (Figure 4).

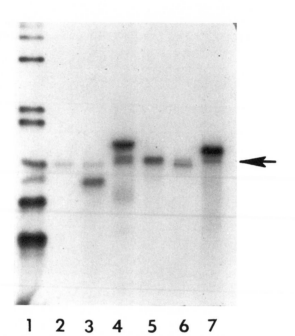

FIG. 4.   Northern blot analysis of mRNA levels for env4 and integrase constructions. Lane 1, glyoxylated, end-labeled DNA standards (HindIII/I, HaeIII/fX174); Lane 2, control yeast cells transformed with pBS24; Lane 3, pBS24 env4; Lane 4, pBS24SODenv4; Lane 5, pBS24Ubenv4; Lane 6, pBS24Integrase; Lane 7, pBS24UbIntegrase. Densitometric scanning gave approximate ratios of message levels of 1:1.5:0.85 for env4, the full length SODenv4 and Ubenv4 messages respectively. Similarly, an approximate ratio of 1:4.5 was observed for integrase and the ubiquitin-integrase fusion messages respectively. The position of the endogenous GAPDH message at approximately 1.3 kbp is shown (arrowed).

Discussion

In this study, we have demonstrated that several regions of the HIV1 genome that have previously been found to be difficult to express, can be produced at extremely high levels in yeast when fused to yeast ubiquitin. As has been shown previously for other proteins *(5,14-17)*, the heterologous protein is quantitatively and precisely cleaved from ubiquitin by an endogenous yeast hydrolase to give a product similar to that produced by direct expression. We have also analyzed the HIV1 products for amino-terminal authenticity. Surprisingly, in contrast to ubiquitin fusion-derived hγ-IFN and α_1-PI, where each amino terminus could potentially be, but is not modified by specific cellular proteolysis pathways *(15)*, the amino-terminal Glu residue of a high proportion of the env4 protein is modified by addition of Arg, to give a heterogeneous final product.

The mechanism for increased expression using the ubiquitin fusion approach is currently unclear. Ecker *et al.* have reported that levels of ubiquitin fusion mRNA are not increased despite up to several hundred-fold increases in protein levels *(17)*. In the present study, we have observed increases in steady-state message levels for the HIV1 integrase fusion that could account for the observed increase in protein production. For env4, however, where protein levels using the ubiquitin vector were considerable higher, fusion messages were even marginally lower than for directly expressed env4. We conclude therefore, as proposed previously *(17)*, that several mechanisms could account for the increased expression observed. These include increased transcription or message stability, increased mRNA translatability, cellular compartmental targeting by the yeast ubiquitin fusion partner, or a combination of these factors. Interestingly, the amino termini of many of the proteins described in the present study are amino acids of the destabilizing class according to the N-end rule *(5,7,8)*. Our results emphasize that the N-end rule pathway is clearly protein specific and indeed, that the spectrum of half-lives of a series of amino-terminally mutated protein analogs could also be related to their levels of expression and their solubility within the cell. For example, we have replaced the amino-terminal Gln residue of hγ-IFN with the highly destabilizing Arg residue and found identical expression levels for each protein. Electron microscopic analysis has shown dense inclusion bodies for hγ-IFN expressed at high levels in yeast *(35)*. In this case, therefore, it seems likely that the rate limiting step in protein degradation in yeast is a function of protein solubility. To test this hypothesis, it would be necessary to analyze amino-terminally mutated hγ-IFN expression under conditions where the expressed protein remained fully soluble and accessible to specific proteolysis. Indeed, these are precisely the conditions under which the N-end rule pathway was determined for amino-terminally mutated bacterial β-galactosidsase and DHFR *(5,7,8)*.

Perhaps also related to the question of solubility, is our observation of high levels of env4 containing an additional Arg residue at its amino terminus. Varshavsky and co-workers *(5,7,8)* had previously shown that β-galactosidase analogs bearing Glu (or Asp) at their amino termini were modified by arginyl-tRNA-protein transferase catalyzed addition of arginine, prior to rapid degradation on the ubiquitin-dependent proteolysis pathway. In contrast however, we found no modification of the amino terminal glutamic acid residue of α_1-PI, expressed at high levels in yeast *(15* and present study). It is possible, that the extremely high levels of fully soluble α_1-PI overload these modification mechanisms and also, that any α_1-PI that becomes modified by addition of Arg is highly susceptible to rapid degradation. In contrast, env4 is highly insoluble in yeast and can only be extracted using strong denaturants. Consistent with our findings therefore, is that env4 with the arginine modification forms inclusion bodies, along with unmodified env4, thereby becoming refractory to proteolytic degradation, and eventually constitutes the major proportion of the purified env4 product. In using the ubiquitin fusion approach for heterologous gene expression it is therefore advisable, where possible, to avoid the secondary destabilizing amino acid residues Glu, Asp and Cys, and perhaps also the tertiary residues Gln and Asn when designing the eventual amino-terminal structure of a recombinant polypeptide.

The availability of large quantities of env4 now allows studies to compare responses of env4 immunized test animals with those immunized with recombinant env1 *(22)*, the non-glycosylated amino-terminal half of gp120 and also env2 *(22)* the non-glycosylated equivalent of gp120. Similarly, comparisons can also be made regarding the ability of each molecule to extract HIV1 neutralizing antibodies from infected individuals when bound to solid supports *(2)*. Comparisons of this type may be critical in designing potential vaccine candidates, and also for the analysis of cross-neutralization of HIV1 isolates from divergent geographical locations *(23,36)*.

Finally, we have shown the ubiquitin fusion approach to be particularly amenable to the production of HIV1 integrase in yeast. The high levels of expression have allowed purification of integrase to homogeneity, and activity studies using synthetic DNA substrates are currently in progress. Recombinant enzymes derived from the *pol* gene of HIV1 are currently under intense study as adjuncts to the rational design of potential chemotherapeutic agents against AIDS. To this end, structural studies on two of the enzymes, the aspartyl protease and RT, have led to their crystallization, and for protease, total determination of its three-dimensional structure *(37,38)*. Also, we have described *in vitro* assay systems for yeast-derived HIV1 protease and RT *(19, 20)*. Integrase, the third enzyme of the HIV1 *pol* gene is now available for similar biochemical and structural studies. For this type of study, the amino-terminal authenticity afforded by the ubiquitin fusion approach is highly desirable for this enzyme, and also perhaps, for other human pathogen-associated enzymes.

Acknowledgments

We thank S.H. Chamberlain and Dr. F.R. Masiarz for protein sequence analysis, O. Mason for DNA sequencing and Toni H. Jones for preparation of the manuscript. We also thank our many friends and colleagues at Chiron for their continued support. This work was supported by Chiron Corporation, and a grant from the California State HIV Vaccine Development Program (to K.S.S.).

Literature Cited

1. Yarden, Y.; Escobedo, J.A.; Kuang, W.J.; Yang-Fen, T.L.; Daniel, T.O.; Tremble, P.M.; Chen, E.Y.; Ander, M.E.; Harkins, R.N.; Francke, U.; Fried, V.A.; Ullrich, A. and Williams, L.T. *Nature* **1986** *323*, 226-232.
2. St. John, T.; Gallatin, W.M.; Siegelman, M.; Smith, H.T.; Fried, V.A. and Weismann, I.L. *Science* **1986** *231*, 845-850.
3. Rechsteiner, M. *Annu. Rev. Cell. Biol.* **1987** 3, 1-30.
4. Busch, H. *Methods Enzymol.* **1984** *106*, 238-262.
5. Bachmair, A.; Finley, D. and Varshavsky, A. *Science* **1986** *234*, 179-186.

6. Finley, D.,; Ozkaynak, E; Jentsch, S.; McGrath, J.P.; Bartel, B.; Pazin, M.; Sapka, R.M. and Varshavsky, A. in *Ubiquitin* (Rechsteiner, M., ed) **1988** pp 39-74, Plenum Press, New York.
7. Varshavsky, A.; Bachmair, A.; Finley, D.; Gonda, D. and Wünning, I. in *Ubiquitin* (Rechsteiner, M., ed) **1988** pp 287-324, Plenum Press, New York.
8. Varshavsky, A.; Bachmair, A.; Finley, D.; Gonda, D. and Wünning, I. in *Yeast Genetic Engineering* (Barr, P.J., Brake, A.J., and Valenzuela, P., eds) **1989** pp 109-143, Butterworths, New York.
9. Finley, D.; Bartel, B. and Varshavsky, A. *Nature* **1989** *338*, 394-401
10. Ozkaynak, E.; Finley, D. and Varshavsky, A. *Nature* **1984** *312*, 663-666.
11. Ozkaynak, E.; Finley, D.; Soloman, M.J. and Varshavsky, A. *EMBO J.* **1987** *6*, 1429-1439.
12. Miller, H.I.; Henzel, W.; Ridgway, J.; Kuang, W.-J.; Chisholm, V. and Liu, C.-C. *Biotechnology* **1989** *7*, 698-704.
13. Monia, B.P.; Ecker, D.J.; Crooke, S.T. *Biotechnology* **1990**, *8*, 209-215.
14. Barr, P.J.; Shuster, J.R.; Bathurst, I.C.; Cousens, L.S.; Lee-Ng, C.T.; Gibson, H.L. and Sabin, E.A. *Yeast* **1988** *4*, S24 (Abstr.)
15. Sabin, E.A.; Lee-Ng, C.T.; Shuster, J.R. and Barr, P.J. *Biotechnology* **1989** *7*, 705-709.
16. Butt, T.R.; Khan, M.I.; Marsh, J.; Ecker, D.J. and Crooke, S.T. *J. Biol. Chem.* **1988** *263*, 16364-16371.
17. Ecker, D.J.; Stadel, J.M.; Butt, T.R.; Marsh, J.A.; Monia, B.P.; Powers, D.A.; Gorman, J.A.; Clark, P.E.; Warren, F.; Shatzmann, A. and Crooke, S.T. *J. Biol. Chem.* **1989** *264*, 7715-7719.
18. Bathurst, I.C.; Chester, N.; Gibson, H.L.; Dennis, A.F.; Steimer, K.S. and Barr, P.J. *J. Virol.* **1989** *63*, 3176-3179.
19. Pichuantes, S.; Babé, L.M.; Barr, P.J. and Craik, C.S. *Proteins: Structure, Function and Genetics* **1989** *6*, 324-337.
20. Barr, P.J.; Power, M. D.; Lee-Ng, C.T.; Gibson, H.L. and Luciw, P.A. *Biotechnology* **1987** *5*, 486-489.
21. Truett, M.A.; Chien, D.Y.; Calarco, T.L.; DiNello, R.K. and Polito, A.J. in *HIV Detection by Genetic Engineering Methods* **1989** (Luciw, P.A., and Steimer, K.S., eds) pp 121-142, Marcel Dekker, New York.
22. Barr, P.J.; Steimer, K.S.; Sabin, E.A.; Parkes, D.; George-Nascimento, C.; Stephans, J.C.; Powers, M. A.; Gyenes, A.; Van Nest, G.A.; Miller, E. T.; Higgins, K.W. and Luciw, P. A. *Vaccine* **1987** *5*, 90-101.
23. Steimer, K.S.; Van Nest, G.A.; Haigwood, N.L.; Tillson, E.M.; George-Nascimento, C.; Barr, P.J. and Dina, D. in *Vaccines 88* **1988** pp 347-355, Cold Spring Harbor, New York.
24. Steimer, K.S.; Higgins, K.W.; Powers, M.A.; Stephans, J.C.; Gyenes, A.; George-Nascimento, C.; Luciw, P.A.; Barr, P.J.; Hallewell, R.A. and Sanchez-Pescador, R. *J. Virol.* **1986** *58*, 9-16.
25. Sanchez-Pescador, R.; Power, M.D.; Barr, P.J.; Steimer, K.S.; Stempien, M.M.; Brown-Shimer, S.L.; Gee, W.; Renard, A.; Randolph, A.; Levy, J.A.; Dina, D. and Luciw, P.A. *Science* **1986** *227*, 484-492.
26. Lightfoote, M.M.; Coligan, J.E.; Folks, T.M.; Fauci, A.S.; Martin, M.A. and Venkatesen, S. *J. Virol.* **1986** *60*, 771-775.
27. Sanger, F.; Nicklen, S. and Coulson, A.R. *Proc. Natl. Acad. Sci. USA* **1977** *74*, 5463-5466.
28. Maniatis, T.; Fritsch, E.F. and Sambrook, J. *Molecular Cloning, A Laboratory Manual,* **1982** Cold Spring Harbor Laboratory, Cold Spring Harbor, New York.
29. Schultz, L.D. *Biochem.* **1978** *17*, 750-758.

30. Shuster, J.R., in *Yeast Genetic Engineering* **1989** (Barr, P.J., Brake, A.J., and Valenzuela, P., eds) pp 83-108, Butterworths, New York.

31. Hinnen, A.; Hicks, J.B. and Fink, G.R. *Proc. Natl. Acad. Sci. USA* **1978** *75*, 1919-1933.
32. Barr, P.J.; Gibson, H.L.; Enea, V.; Arnot, D.E.; Hollingdale, M.R. and Nussenzweig, V. *J. Exp. Med.* **1987** *165*, 1160-1171.
33. LeGendre, N. and Matsudaira, P. *Biotechniques* **1988** 6, 154-159.
34. Arfin, S.M. and Bradshaw, R.A. *Biochemistry* **1988** *27*, 7979-7984.
35. Van Brunt, J. *Biotechnology* **1986** *4*, 1057-1062.
36. Haigwood, N.L.; Shuster, J.R.; Moore, G.K.; Mann, K.A.; Barker, C.B.; Lee, H.; Pruyne, P.T.; Barr, P.J.; George-Nascimento, C.; Scandella, C.J.; Skiles, P.; Higgins, K.W. and Steimer, K.S. *AIDS* Res. Hum. Retrovir., **1990** *6*, 855-869.
37. Navia, M.A.; Fitzerald, M.D.; McKeever, B.M.; Leu, C-T.; Heimbach, J.C.; Herber, W.K.; Sigal, I.S.; Darke, P.L. and Springer, J.P. *Nature* **1989** *337*, 615-620.
38. Lowe, D.M.; Aitken, A.; Bradley, C.; Darby, G.K.; Larder, B.A.; Powell, K.L.; Purifoy, D.J.M.; Tisdale, M. and Stammers, D.K. *Biochemistry* **1988** *27*, 8884-8889.

RECEIVED January 16, 1991

Chapter 6

Constitutive and Regulated Expression of the Hepatitis B Virus (HBV) PreS2+S Protein in Recombinant Yeast

P. J. Kniskern, A. Hagopian, D. L. Montgomery, C. E. Carty,
P. Burke, C. A. Schulman, K. J. Hofmann, F. J. Bailey, N. R. Dunn,
L. D. Schultz, W. M. Hurni, W. J. Miller, R. W. Ellis, and R. Z. Maigetter

Merck Sharp and Dohme Research Laboratories, West Point, PA 19486

Recombinant strains of Saccharomyces cerevisiae,
expressing the PreS2+S envelope protein of the
hepatitis B virus (HBV), were constructed using
either the constitutive glyceraldehyde-3-phosphate
dehydrogenase (GAP) promoter or the regulated
galactose-1-phosphate uridyltransferase (GAL10)
promoter. These strains were compared by determining not only the yield of PreS2+S antigen but also
the retention of the plasmid during growth and
expression. When the GAP system was examined, the
expression of PreS2+S was high, but plasmid levels
decreased significantly during growth. For the GAL10
system, the expression of the PreS2+S was significantly lower, but the plasmid copy number remained
high. Much higher expression levels (comparable to
those obtained for GAP promoter) were achieved with
the GAL10 promoter when PreS2+S was expressed in a
host strain containing a chromosomally integrated
expression cassette comprised of the GAL10 promoter
fused to the GAL4 positive regulatory gene, and the
plasmid level remained high.

Hepatitis B virus (HBV) is the infectious agent responsible for
several varieties of human liver disease. Many infected individuals suffer through an acute phase of disease which is followed
by recovery. However, a number of individuals fail to clear their
infection, thereby becoming chronic carriers of the virus. HBV
infection is endemic in many parts of the world, with a high
incidence of transmission occurring perinatally from chronically
infected mothers to their newborns who themselves often become
chronically infected. The number of chronic carriers worldwide
has been estimated at over two hundred and fifty million, and

0097–6156/91/0477–0065$06.00/0
© 1991 American Chemical Society

hundreds of thousands die annually from cirrhosis and/or hepato-
cellular carcinoma, the long-term consequences of chronic
hepatitis B (1-3).

The development of vaccines for the prevention of HBV disease
has occupied the research efforts of numerous laboratories during
the past two decades. These efforts led to a first generation
vaccine comprised of the 22 nm particle or HBV surface antigen
(HBsAg or S) particles which were purified from the plasma of
chronic carriers (4). This vaccine (HEPTAVAX-B) has been available
since 1981 and has proven to be both safe and highly efficacious
(5-7). However, even as this vaccine was being developed, concern
arose that the quantities of high-titered plasma from asymptomatic
carriers might be insufficient for production of enough vaccine for
worldwide use. Therefore, researchers turned to recombinant DNA
technology to express the HBV envelope proteins. Thus, HBsAg was
expressed in a recombinant strain of S. cerevisiae, purified and
formulated into a second-generation hepatitis B (HB) vaccine (8-10).
This vaccine (RECOMBIVAX HB) which, like HEPTAVAX-B is composed of
22 nm particles containing HBsAg, was licensed in 1986 and, like the
first-generation HEPTAVAX-B, is relatively safe, well-tolerated and
highly efficacious (11-13).

Although both HEPTAVAX-B and RECOMBIVAX HB as well as other
licensed vaccines expressed in yeast (such as ENGERIX-B and
BIMMUGEN) contain exclusively the S domain (226 AA), the envelope
proteins of HBV are the translational products of a large viral
open reading frame (ORF) encoding 389 amino acids (aa). This ORF is
demarcated into three domains, each beginning with an ATG codon
capable of functioning as a translational initiation site. These
domains define three polypeptides referred to as S or HBsAg (226
aa), PreS2+S (281 aa) and PreS1+PreS2+S (389 aa), also referred to
as: p24/gp27, p30/gp33/gp36 and p39/gp42, respectively (14-15).
Recently, several independent lines of evidence have suggested that
the PreS sequences may be important in immunity to HBV (16-22).

In light of these observations and because of the utility of
recombinant yeast in producing HB vaccines (9,23-24), many groups
have turned their attention to the development of yeast-derived HB
vaccine candidates containing not only the already proven efficacious S domain but also all or part of the PreS domain (10,25).

The yeast-derived HB vaccines successfully developed to date
as commercial vaccines depend on strong constitutively active
promoters for expression of the HBsAg polypeptide (26-27).
However, constitutive expression of the PreS1+PreS2+S polypeptide,
is toxic for yeast and this polypeptide could be expressed successfully only by using a regulatable promoter system (25). The
situation for the toxicity of the PreS2+S polypeptide is intermediate between the cases for the expression of PreS1+PreS2+S and
the expression of S only. The PreS2+S peptide has been expressed
to high levels with the GAP promoter (10), but this constitutive
expression was achieved in the face of significant toxicity to the
yeast which manifested itself in a decreased growth rate and
significant loss of plasmid during growth (data not shown).

Because of the successful regulated expression of PreS1+PreS2+S and other polypeptides with toxicity for yeast such as the Epstein-Barr virus (EBV) gp350 (28), it was reasoned that the expression of the PreS2+S peptide under physiologic regulation might result in improved growth, plasmid retention and good antigen productivity. The inducible galactose-1-phosphate uridyltransferase or GAL10 promoter (pGAL10), used for the successful expression of EBV gp350, has the additional advantage of being up-regulated in a host strain containing an integrated expression cassette comprising pGAL10 fused to the GAL4 positive regulatory gene (29). This monograph thus addresses studies comparing growth characteristics, plasmid retention and antigen productivity in shake flasks and fermentors for the constitutive pGAP system and systems employing the regulatable pGAL10.

Materials and Methods

Recombinant DNA.   Reagents, techniques, and plasmids for molecular cloning and expression have been described previously (28,30).

Expression Plasmids.   The expression plasmid pHBpreSGAP347/19T (Figure 1a) was obtained from P. Valenzuela and has been described previously (31). A second construction also employing the constitutive glyceraldehyde-3-phosphate dehydrogenase promoter (pGAP) was engineered to lack any viral flanking sequences (10) and is designated pYGpreS2S-1 (Figure 1b). For studies of physiologically regulated expression, the pGAP promoter from pYGpreS2S-1 was replaced with pGAL10 creating the galactose inducible construction designated pYGAL10preS2S-1 (Figure 1c).

Yeast Strains.   Strains 2150-2-3 (MATa, ade1, leu2-04, cir^o) and CF42 (MATa/$\propto$, ade1, leu2-04, ura3, cir^o) have been described previously (10,31). Strain Sc252 (pGAL10-GAL4) [MATa, ura3-52, his3:GAL10p-GAL4-URA3, leu2-2,112, ade1, cir^o], a strain containing an integrated pGAL10-GAL4 expression cassette which overproduces the GAL4 gene product, has been described elsewhere (29).

Expression Clones.   The expression clones in Table I were established by transforming the yeast strains with the expression plasmids by standard techniques (28,30).

Table I. Constructions of Expression Clones

Plasmid	Promoter	Yeast Strain	GAL4 Phenotype	Transformant
pHBpreSGAP347/19T	GAP	2150-2-3	wild type	GAP1
pYGpreS2S-1	GAP	CF42	" "	GAP2
pYGAL10preS2S-1	GAL	CF42	" "	GAL1
pYGAL10preS2S-1	GAL	SC252 (pGAL10-GAL4)	overproducer	GAL2

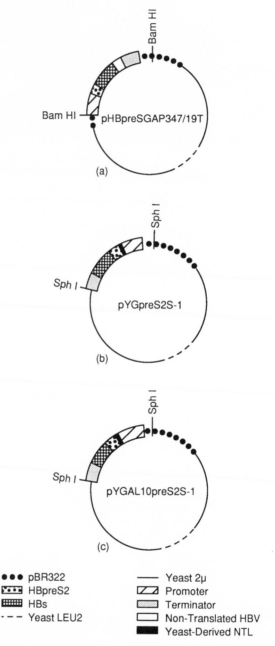

Figure 1.   Plasmid constructions for expression, in yeast, of:
(a) PreS2+S with the GAP promoter designated pHBpreSGAP347/19T;
(b) PreS2+S without any viral flanking sequence and using the
GAP promoter designated pYGpreS2S-1; and (c) PreS2+S with the
GAL10 promoter designated pYGAL10preS2S-1.

<u>Preparation of Subcultures</u>. Subcultures of the clones were prepared
by streaking for isolated colonies from transformation plates.
These clones subsequently were grown to confluence on leucine minus
(Leu⁻) agar (<u>32</u>) at 28°C. The cells were harvested into yeast
extract, soy peptone and glucose (YEHD medium, <u>32</u>) containing 17%
glycerol, and 1-mL aliquots were frozen immediately at -70°C; these
were designated premaster cultures. For preparation of a master
stock, a premaster vial was thawed and plated onto Leu⁻ agar. After
3 days, at 28°C, the cells were removed and frozen vials prepared as
described above. For preparation of working stock seed, a master
seed vial was thawed and expanded by the same protocol.

<u>Fermentation Studies</u>. Cells were streaked onto Leu⁻ agar and grown
at 28°C. After two to three days, cells were harvested, inoculated
into non-baffled 250-mL shake flasks containing 50 mL YEHD or a Leu⁻
medium and placed on an orbital shaker at 350 rpm at 28°C.
    For shake flask studies with the GAP cultures, after the cells
were grown in either Leu⁻ or YEHD in non-baffled 250-mL shake flasks
for 12-15 hrs, a 0.25% (v/v) inoculum was placed into either a
non-baffled 250-mL flask with 50 mL of YEHD medium or a non-baffled
2-L flask with 500 mL of YEHD; the cells were grown at 350 rpm at
28°C for 40-48 hrs and harvested. If an additional seed stage was
used, a 0.25% (v/v) inoculum would be transferred to a non-baffled
250-L flask with 50 mL YEHD and grown for 12-15 hrs at 28°C.
    For shake flask studies with the GAL cultures, when cells
reached mid-exponential phase of growth in non-baffled 250-mL shake
flasks, a 0.25% (v/v) inoculum was introduced into a 2-L non-baffled
shake flask containing 500 mL of YEHD.
    When the concentration of glucose was approximately 2 g/L,
galactose was added to a final concentration of 20 g/L to induce
expression of PreS2+S. The fermentation was continued for an additional 40-48 hours after galactose addition (<u>33</u>).
    When fermentors were used, cells were harvested from Leu⁻
plates and inoculated into non-baffled 2-L shake flasks containing
500 mL YEHD. The 2-L shake flasks were grown at 28°C and 350 rpm
for 18 hr, from which a 0.25% (v/v) inoculum was placed into either
a 16-L or a 250-L fermentor. The 16-L fermentor was operated at
28°C, 500 rpm and 5 L/min aeration with pH controlled at 7.0 using
2 N NaOH and 9 N H_3PO_4; the 250-L fermentor was operated at 28°C,
150 rpm, 60 L/min aeration with similar pH control. As demonstrated
for shake flasks, galactose was added when the concentration of
glucose was approximately 2 g/L. The fermentations were harvested
and assayed after 40-48 hrs (<u>33</u>).

<u>Analytical Methods</u>. Glucose concentrations and replica plating
assays were performed as described (<u>32</u>). The copy number assay was
similar to that described previously (<u>34</u>); the samples were digested
with <u>Eco</u>RI, run on agarose gels, stained with ethidium bromide and
the plasmid band quantitated relative to a ribosomal band using
scanning densitometry. The analysis of PreS2+S was performed as
outlined previously (<u>10</u>). Briefly, PreS2+S was quantitated by a
radioimmunoassay (PreS RIA) based on the affinity of the PreS
protein for polymerized human serum albumin.

A pellet of yeast cells was resuspended in phosphate buffer with
2 mM PMSF, 0.1% Triton X-100, glass beads, broken on a Big Vortexer
[GLAS-COL, Terre Haute, IN] and clarified by centrifugation ($\underline{30}$).
Polystyrene beads coated with polymerized human serum albumin were
incubated with samples of supernatant fluid from the broken yeast
cells and with known standards. The beads were washed, incubated
with [125]-I labeled antibody to HBsAg, washed and counted. Samples
were quantitated relative to a standard curve and expressed as PreS
RIA units. The supernatant fluid of the broken cell extract also
was assayed for protein content ($\underline{35}$). The expression of PreS2+S is
expressed relative to a GAP1 premaster culture that was run as
internal standard in each experiment.

Results

pGAP Expression Systems. Since the pGAP was used for expression of
S polypeptides ($\underline{31},\underline{32}$), this expression system was selected for
initial studies with PreS2+S. Shake flasks were used with the
premaster of the GAP1 cultures to determine the conditions for
scale-up of the fermentation process. When a 0.25% or 5% (v/v)
inoculum was used, there was no significant effect on the relative
antigen productivity (data not shown). Therefore, because a 0.25%
inoculum would eliminate the need for an intermediate fermentor,
this inoculum was selected for scale-up experiments. When this
inoculum size was used for fermentors, the relative productivity
was comparable to the values obtained in shake flasks.
    In order to establish seeds for expanded studies in fermentors,
the premaster vial was subcultured to prepare large numbers of
master and working stock cultures. When these subcultures were
tested in shake flasks, there was approximately a 10-fold loss in
activity at the working stock stage (Table II) as compared to the
master and premaster cultures. Therefore, subsequent experiments
were performed with cultures prepared from the master seed rather
than the working stock seed.

Table II.   Production of HBV PreS2+S upon subculturing

Culture designation	Relative expression of PreS2+S*		
	premaster	master	working
GAP1	1.0	1.0	0.1
GAP2	5.0	7.0	0.2
GAL1	1.0	1.0	1.0
GAL2	5.0	5.0	5.0

*Antigen productivity is expressed relative to a GAP1
(premaster) culture run as an internal standard in each
experiment.

Although the relative expression of PreS2+S of cultures from
the master seed was adequate for further development, an increased
level of expression was desirable as this would provide a higher
purity starting material and potentially an improved vaccine ($\underline{36}$);

it also was hoped that this would produce a culture which could be passaged to make a productive working stock seed. Previous research in our laboratory had shown higher expression of the hepatitis B core antigen gene in yeast transformed with plasmids in which yeast rather than viral DNA sequences flanked the gene (30). Based upon this strategy, an analogous construction for PreS2+S was engineered (Figure 1b). Yeast transformed with this vector (pYGpreS2S-1) and designated GAP2 (Table I) were established as premaster cultures and tested in the shake flasks. The relative expression was nearly five-fold higher for this strain than for the GAP1 strain (Table II).

Although the GAP2 construct produced more PreS2+S than GAP1, two concerns remained. First, when GAP2 was examined at the end of the fermentation (Table III), there was a same high loss of the Leu[+] phenotype which also had been observed for GAP1 and which ranged from 30-70%. Second, when subcultures were prepared from the master seed of GAP2, there was a similar loss in the relative expression of PreS2+S at the working stock seed which had been observed for GAP1 (Table II). Furthermore, when samples were examined for plasmid copy number, the seed culture grown in Leu[-] broth had ca. 100 copies per cell; after 48 hrs of growth in fermentors in YEHD broth, the plasmid was barely detectable (data not shown). Interestingly, as shown in Table III, productivity remained high even under these conditions of apparently nearly complete plasmid loss and further examination of six sibling clones of GAP2 showed that two clones produced high and consistent yields of PreS2+S following multiple passages of the culture (data not shown).

Table III. Cultivation of the GAP2 culture
compared to GAL2 culture

Transformant	Relative Expression of PreS2+S*	% Retention of Leu[+] Phenotype
GAP2	5	33-77
GAL2	5	98-100

*Antigen productivity is expressed relative to a GAP1 (premaster) culture run as an internal standard in each experiment.

Regulated expression systems (pGAL10). Having demonstrated a high expression of PreS2+S with GAP2, a parallel research effort was initiated using a regulated promoter. This effort was begun to minimize plasmid loss and the potential for decreased antigen productivity especially at the working stock seed level. Based upon the earlier studies with the galactose system for regulated expression of EBV gp350 (28), expression plasmids were engineered with this promoter (Figure 1c). This plasmid is identical to that developed for the GAP2 except that pGAL10 was used in place of pGAP to express the PreS2+S antigen.

When GAL1 (Table I) was examined using either Leu or YEHD medium in the seed train and YEHD medium in the production flask, the expression of PreS2+S was independent of seed medium composition (data not shown). Also, under these conditions, there was good cell growth and virtually no loss of the Leu[+] phenotype.

When GAL1 was subcultured to prepare master and working stock seeds, there was no loss of antigen productivity (Table II). However, the yield of PreS2+S was only ca. 20% of the value obtained for GAP2 (Table II). Because another advantage of employing the inducible GAL10 promoter is the potential for up-regulation in strains which overproduce the positive regulatory GAL4 gene product (29), the plasmid pYGAL10PreS2S-1 was transformed into a yeast strain that contained the GAL4 gene driven by pGAL10 and integrated into the host chromosome to create the recombinant designated GAL2 (Table I).

When the integrant pGAL10 clone (GAL2) was compared to the non-integrant pGAL10 clone (GAL1), there was a five-fold improvement in relative productivity (Table II) to a level that was equivalent to GAP2. Furthermore, GAL2 could be subcultured up to at least 7 times without a detectable loss in relative expression of PreS2+S or Leu$^+$ phenotype (Table III). In addition, when samples were examined for plasmid copy number, there were approximately 300 copies of plasmid/cell when cells were maintained on Leu$^-$ medium. After growth in YEHD, there were 100-150 copies of plasmid/cell both prior to induction and at the end of the fermentation.

Discussion

This paper explores several approaches in molecular biology and in fermentation which enhance the expression of the HBV PreS2+S protein in recombinant yeast. Our results show that the PreS2+S expression levels can be increased at the molecular level by using yeast-derived rather than virus-derived DNA sequences to flank the PreS2+S gene and by exploiting pGAL10 expression plasmids in hosts containing an integrated positive regulatory GAL4 gene. The first clone (GAP1) used for these studies contained viral, nontranslated DNA flanking the PreS2+S ORF sequences. The yield of PreS2+S for this culture was low. A construction (GAP2) containing only yeast-derived non-translated flanking sequences was five-fold more productive than GAP1. However, both GAP1 and GAP2 cultures showed a significant loss in the number of plasmids upon subculturing or fermentation. The reason for this loss may be the toxic nature of preS2 for yeast cells. Previous studies with PreS1+PreS2+S had shown that yeast cells transformed with a plasmid constitutively expressing PreS1+PreS2+S failed to recover on selective agar plates even after seven days of incubation (25). For a similar S plasmid, transformants are observed in 3-4 days and constitutive PreS2+S transformants require 4-5 days for recovery. In addition, subculturing the pGAP clones resulted in a loss of antigen expression after two passages from the original clone, which also suggested a yeast-inhibitory effect for the PreS2+S protein. To eliminate this decline in antigen expression associated with subculturing the pGAP clones, we examined sibling clones and checked their expression levels through several subcultures. Approximately one-third of the sibling clones could be subcultured without loss of antigen productivity; these clones may have adapted to the expression of PreS2+S and were more resistant to "toxic" effects of this protein. Although we continue to study constitutive expression of PreS2+S, these potential unstable events made it apparent that we should also examine systems wherein the "toxic" protein is expressed only after the cells had been grown to high-cell densities.

We have demonstrated through optimal vector construction, clonal selection, and utilization of inducible expression systems in tailored host cells, that such recombinant yeast strains expressing the PreS2+S polypeptide have been established. The availability of such recombinant yeasts may allow the development of a new HBV vaccine with perhaps greater breadth or duration of protection than that obtained for the current highly efficacious vaccines.

Acknowledgments

We thank J. Broach, J. Hopper, J. Tomassini, and P. Valenzuela for some of the yeast strains, plasmids and DNA used in these studies; A. Bertland, M. Johnston, B. Rich, R. Roehm, E. Wasmuth, and A. Wolfe for analytical testing; J. Broach, A. Conley, S. Drew, R. Gerety, W. Herber, J. Hopper, and W. Miller for helpful discussions; R. Zeigler for artwork and G. Albanesius and D. Sagel for careful preparation of the manuscript.

Literature Cited

1. Szmuness, W. Prog. Med. Vir. 1978, 24, 40-69.
2. Szmuness, W.; Harley, E.J.; Ikram, H.; Stevens, C.E. In Viral Hepatitis; (Vyas, G.N.; Cohen, S.N. and Schmid, R., Eds., Franklin Institute: Philadelphia 1978; pp. 297-320.
3. Beasley, R.P.; Hwang, L.-Y. In Viral Hepatitis and Liver Disease. Vyas, G.N.; Dienstag, J.H.; Hoofnagle, J.H., Eds.; Grune & Stratton: Orlando, 1984; pp. 209-224.
4. Hilleman, M.R.; McAleer, W.J.; Buynak, E.B.; McLean, A.A. J. Infect. 1983, 7, 3-8.
5. Szmuness, W.; Stevens, C.E.; Zang, E.A.; Harley, E.J.; Kellner Hepatology 1981, 1, 377-385.
6. Tabor, E.; Buynak, E.B.; Smallwood, L.A.; Snoy, P.; Hilleman, M.R.; Gerety, R.J. J. Med. Virol. 1983, 11, 1-10.
7. Francis, D.P.; Feorino, P.M.; McDougal, S.; Warfield, D.; Getchell, J.; Cabradilla, C.; Tong, M.; Miller, W.J.; Schultz, L.D.; Bailey, F.J.; McAleer, W.J.; Scolnick, E.M.; Ellis, R.W. J. Am. Med. Assn. 1986, 256, 869-872.
8. McAleer, W.J.; Buynak, E.B.; Maigetter, R.Z.; Wampler, D.E.; Miller, W.J.; Hilleman, M.R. Nature 1984, 307, 178-180.
9. Kniskern, P.J.; Hagopian, A.; Burke, P.; Dunn, N.; Montgomery, D.L.; Schultz, L.D.; Schulman, C.A.; Carty, C.E.; Maigetter, R.Z.; Wampler, D.E.; Lehman, E.D.; Yamazaki, S.; Kubek, D.J.; Emini, E.A.; Miller, W.J.; Hurni, W.M.; Ellis R.W. In Immunobiology of Proteins and Peptides V; Atassi, M.Z.; Ed.; Plenum Pub. Corp.
10. Ellis, R.W.; Kniskern, P.J.; Hagopian, A.; Schultz, L.D.; Montgomery, D.L.; Maigetter, R.Z.; Wampler, D.E.; Emini, E.A.; Wolanski, B.; McAleer, W.J.; Hurni, W.M.; Miller, W.J. In Viral Hepatitis and Liver Disease; Zuckerman, A.J., Ed.; Alan R. Liss, New York, 1988; pp. 1079-1086.

11. Stevens, C.E.; Alter, H.J.; Taylor, P.E.; Zaug, E.A.; Harley, E.J.; Szmuness, W. N. Engl. J. Med. 1984, 311, 496-501.
12. Zajac, B.A.; West, D.J.; McAleer, W.J.; Scolnick, E.M. J. Infect. 1986, 13(A), 39-45.
13. Stevens, C.E.; Taylor, P.E.; Tong, M.J.; Toy, P.T.; Vyas, G.N.; Krugman, S. J. Am. Med. Assoc. 1987, 257, 2612-2616.
14. Heermann, K.H.; Goldmann, U.; Schwartz, W.; Seyffarth, T.; Baumgarten, H.; Gerlich, W.H. J. of Virol. 1984, 52, 396-402.
15. Tiollais, P.; Pourcel, C.; Dejean, A. Nature (London) 1985, 489-495.
16. Neurath, A.R.; Kent, S.B.; Strick, N.; Stark, D.; Sproul, P. J. Med. Virol. 1985, 17, 119-25.
17. Neurath, A.R.; Kent, S.B.; Parker, K.; Prince, A.M.; Strick, N.; Brotman, B.; Sproul, P. Vaccine 1986, 4, 35-37.
18. Neurath, A.R.; Kent, S.B.H.; Strick, N.; Parker, K.; Courouce, A.-M.; Riottot, M.M.; Petit, M.A.; Budkowska, A.; Girard, M.; Pillot, J. Molecular Immunol. 1987, 24, 975-980.
19. Milich, D.R.; McNamara, M.K.; McLachlan, A.; Thornton, G.B.; Chisari, F.V. Proc. Natl. Acad. Sci. USA 1985, 82, 8168-172.
20. Milich, D.R.; McLachlan, A.; Chisari, F.V.; Kent, S.B.; Thorton, G.B. J. Immunol. 1986, 137, 315-322.
21. Itoh, Y.; Takai, E.; Ohnuma, H.; Kitajima, H.; Tsuda, F.; Machida, A.; Mishiro, S.; Nakamura, T.; Miyakaway, Y.; Mayumi, M. Proc. Natl. Acad. Sci. USA, 1986, 83, 9174-9178.
22. Emini, E.A.; Larson, V.; Eichberg, J.; Conard, P.; Garsky, V.M.; Lee, D.R.; Ellis, R.W.; Miller, W.J.; Anderson, C.A.; Gerety, R.J. J. Med. Virol. 1989, 28, 7-12.
23. Hilleman, M.R. and Ellis, R.W. Vaccine 1986, 4, 75-76.
24. Ellis, R.W., (1989), Recombinant yeast-derived hepatitis B vaccine - the prototype for biotechnologically-derived old vaccines, in press: Scientific and Regulatory Aspects of Biotechnologically-Produced Medical Agents, A Practical Hand Book. Chiu, Y.-Y. H.; Gueriguian, J.L., Eds.; Marcel & Dekker: New York (to be published in 1991).
25. Kniskern, P.J.; Hagopian, A.; Burke, P.; Dunn, N.R.; Emini, E.A.; Miller, W.J.; Yamazaki, S.; Ellis, R.W. Hepatology 1988, 8, 82-87.
26. Valenzuela, P.; Medina, A.; Rutter, W.J.; Ammerer, G.; Hall, B.D. Nature 1982, 298, 347-352.
27. Rutgers, T.; Cabezon, T.; Harford, N.; Vanderbrugge, D.; Descurieux, M.; Van Opstal, O.; Van Vijnendaele, F.; Hauser, P.; Yoet, P.; DeWilde, M. In Viral Hepatitis and Liver Disease, Zuckerman, A.J., Ed.; Alan R. Liss: New York, 1988; 304-308.
28. Schultz, L.D.; Tanner, J.; Hofmann, K.J.; Emini, E.A.; Condra, J.H.; Jones, R.E.; Kieff, E.; Ellis, R.W. Gene 1987, 54, 113-123.
29. Schultz, L.D.; Hofmann, K.J.; Mylin, L.M.; Montgomery, D.L.; Ellis, R.W.; Hopper, J.E. Gene 1987b, 61, 121-133.
30. Kniskern, P.J.; Hagopian A.; Montgomery, D.L.; Burke, P.; Dunn, N.R.; Hofmann, K.J.; Miller, W.J.; Ellis, R.W. Gene 1986, 46, 135-141.

31. Valenzuela, P.; Coit, D.; Kuo, C.H. <u>Biotechnology</u> 1985, 317-320.
32. Carty, C.E.; Kovach, F.X.; McAleer, W.J.; Maigetter, R.Z. <u>J. Ind. Microbiol</u>. 1987, <u>2</u>, 117-121.
33. Carty, C.E.; Montgomery, D.L.; Hagopian, A.; Burke, P.; Dunn, N.; Schultz, L.D.; Kniskern, P.J.; Maigetter, R.Z.; Ellis, R. W. <u>Proceedings Society of Industr. Microbiol</u>. 1988, p. 15.
34. Nasmyth, K.A.; Reed, S.I. <u>Proc. Natl. Acad. Sci.</u> 1980, USA <u>77</u>, 2119-2123.
35. Lowry, O.H.; Rosebrough, N.J.; Farr, A.L.; Randall, R.J. <u>J. Biol. Chem.</u> 1951, <u>193</u>, 265.
36. West, D.J.; Zajac, B.A.; Ellis, R.W.; Gerety, R.J. <u>Infektionsklinik</u> 1988, <u>1</u>, 1-7.
37. Johnston, S.A. and Hopper, J.E. <u>Proc. Natl. Acad. Sci.</u> 1982, USA <u>79</u>, 6971-6975.
38. Laughon, A.; Driscoll, R.; Wills, N.; Gesteland, R.F. <u>Mol. Cell. Biol.</u> 1984, <u>4</u>, 268-275.
39. Johnston, S.A.; Zavortink, M.J.; Debouck, C.; Hopper, J.E. <u>Proc. Natl. Acad. Sci.</u> USA, 1986 <u>83</u>, 6553-6557.

RECEIVED January 16, 1991

Chapter 7

Engineering Studies of Protein Secretion in Recombinant *Saccharomyces cerevisiae*

Mark R. Marten and Jin-Ho Seo

School of Chemical Engineering, Purdue University,
West Lafayette, IN 47907

A recombinant model system consisting of either the *MF*α1 (α-factor) or the *SUC*2 promoter/signal-sequence fused to the *SUC*2 structural gene (codes for the yeast enzyme invertase) was used to study secreted protein localization and kinetics in the yeast *Saccharomyces cerevisiae*. The *MF*α1 and *SUC*2-directed secretion transported similar amounts of total specific invertase activity out of the cytoplasm. Multiple copies of the *SUC*2 gene on a 2μm based plasmid resulted in a four fold increase in total invertase produced over the wild type, and a five fold increase in invertase secreted to the medium. Derepression of the *SUC*2 promoter showed rapid expression and secretion of invertase into the periplasm, with slow passage through the cell wall to the fermentation broth.

With the aid of recombinant DNA technology, foreign genes can now be cloned into a wide variety of host microorganisms so as to overproduce valuable product proteins. Although some of these products have found their way to the market place, there still remains much room for the improvement and optimization of the large scale commercial production of these proteins. *Saccharomyces cerevisiae*, or baker's yeast, is an attractive host for the production of recombinant proteins, as it offers many practical advantages over other microorganisms such as *E. coli* and *B. subtilis*. Yeast contain no endotoxins and the Food and Drug Administration (FDA) has classified the yeast *S. cerevisiae* as "generally regarded as safe" (GRAS) for human consumption. As a result, FDA approval for products expressed in yeast should be much easier than for those produced in bacteria. Large scale production technology and down stream processing operations have been developed in the baking and brewing industries that could be applied to recombinant protein production. Many inexpensive substrates are available for growth of this single-cell microorganism. Furthermore, yeast lend themselves to genetic manipulation and are in many ways similar to higher eucaryotic cells.

The secretion of heterologous proteins from recombinant cells has emerged in recent years as an important aspect of biotechnology. Secretion can also prevent

0097–6156/91/0477–0077$06.00/0

several problems associated with the overproduction of foreign gene products. Accumulated protein may aggregate and then precipitate (as inclusion bodies) requiring unfolding and refolding processes. Native proteases in the host may act on the accumulated foreign product and degrade much of it. Foreign protein accumulation may be harmful to host cell growth and plasmid stability, leading to a decrease in overall process productivity. On a larger scale, many unit-processes involving the separation of intracellular proteins can be avoided when extracellular enzymes are produced. These processes include disruption of cells, nucleic acid removal and several precipitation steps. For biological activity and stability many proteins require post-translational modifications including glycosylation and disulfide bond formation. Glycosylation, or addition of oligosaccharide side chains, occurs when proteins are directed to the endoplasmic reticulum (ER) and Golgi body (GB) of a eucaryotic cell through the secretion pathway. Thus secretion can be fundamental in obtaining biologically active proteins.

The secretion of heterologous proteins from recombinant *S. cerevisiae* offers great promise in improving product yields in industrial fermentations, yet little study has been done on the effect of various genetic and environmental parameters on the kinetics and efficiency of the secretion process. The effect of these factors needs to be characterized under both diverse and dynamic conditions so that the experimental information can be implemented in the future to design recombinant cell fermentation processes with high productivity.

Secretion Pathway

In *S. cerevisiae* only a small number of proteins are actually secreted out of the cytoplasmic space (1). Those proteins most widely studied include invertase, acid phosphotase, α-factor, and type 1 killer toxin (2). Invertase and acid phosphotase are secreted into the periplasmic space (between the plasma membrane and cell wall) or are retained in the cell wall (3 ,4). α-Factor and type 1 killer toxin pass through the cell wall and are secreted into the growth medium (5 ,6).

The following information is a composite of results from various secretion systems (including yeast). In the first step of the yeast secretion pathway, a gene coding for a secreted product is transcribed and the message is transported to the cytoplasm of the cell where it binds to the two ribosomal subunits. After the first 60-70 amino acids are translated, a short hydrophobic segment of between 20-30 residues termed the signal sequence emerges from the ribosome. Free "signal recognition particles" (SRP) in the cytoplasm then bind the nascent signal sequence and arrest further translation of the protein. The SRP directs the entire complex to the membrane of the rough endoplasmic reticulum (RER) where the SRP binds to integral membrane "docking proteins" (DP), releasing arrest of translation and allowing concomitant translocation of the polypeptide through the membrane into the lumen of the RER (7 ,8). Recently several groups working with *in vitro* yeast translocation systems have suggested the necessity of several other gene products for transport into the RER (9 ,10). Some of these belong to the same family as the HSP70 gene products which are thought to be unfoldases inhibiting three-dimensional conformation of nascent proteins thus enhancing their ability to move through the RER membrane (11).

The post-translational events in the yeast secretory pathway have been determined genetically using temperature-sensitive secretory *(sec)* mutants (12). These cells grow normally at the permissive temperature of 25°C but, when raised to the nonpermissive temperature of 37°C fail to export normally secreted proteins (1). A series of double *sec* mutants was developed that led to the elucidation of much of the post-translational yeast secretion pathway (13) , as extensively reviewed by Pfeffer (14). For several reasons, many researchers have used the naturally secreted yeast enzyme invertase as a model protein for study of the secretion pathway (15 ,16). It was found that as invertase enters the RER, the hydrophobic amino terminal residues are removed by a lumenal signal peptidase (associated with the RER membrane) and the initial steps of glycosylation occur (17). Core oligosaccharide is assembled on a phosphorylated dolichol and involves the stepwise addition of 14 sugars including glucose, mannose, and *N*-acetylglucoseamine residues (11). Eleven genes are required for proper assembly and modification of these complexes which are eventually transferred to the polypeptide by an oligosaccharide-transferase at an asparagine residue which is followed by any amino acid then either serine or threonine (18). Nine or more *sec* gene products and energy in the form of ATP are then required to transfer the protein to the Golgi body (GB) (13) where segregation from proteins targeted to the lysosome is thought to take place (19). In the GB the oligosaccharides are elongated by addition of outer-chain carbohydrate (15). New evidence suggests that the yeast Golgi is divided into functionally different compartments which specialize in the addition of outter chain oligosaccharide determinants (20). Two or more *sec* gene products and energy are required to package the fully glycosylated proteins into secretion vesicles (13). With the aid of flourecien-labeled invertase antibodies it was shown that the secretion vesicles are then transported into the budding yeast daughter cell to release their contents (4). Similar behavior was also observed with the native yeast protein acid phosphatase (21 ,22). The secretion vesicles fuse with the plasma membrane of the daughter cell in a process that requires at least ten additional *sec* gene products and energy (13). The contents of the vesicle are then released from the cell into the periplasm. The secretion pathway in yeast is characterized by low levels of protein precursors and few secretory organelles (23)

The mechanisms for partitioning of the secreted protein between the cell wall and the extracellular medium is not as yet clear. In many wild type strains, the bulk of both invertase and acid phosphatase is retained in the cell wall or the periplasmic space(3 ,4). Although both of these are glycoproteins of quite high molecular weight, these properties do not of themselves prevent polypeptides from crossing the cell wall. Julius et al. (24) isolated the *KEX2* gene which codes for the endopeptidase responsible for initial precursor cleavage of the *MFα1* gene-product α-factor. Strains with a Kex2⁻ genotype secreted into the culture medium hyperglycosylated prepro α-factor which had a molecular weight of approximately 150 kdal. Carter et al. (25) obtained similar results when the entire *MF*α1 promoter and prepro portions were fused to β-interferon and the cloned gene was introduced into Kex2⁻ strains by transformation. The fusion product had a molecular weight of over 100 kdal and yet was found in the culture medium. These results would indicate that the yeast cell wall does not act as a molecular sieve. To date, no simple model exists to explain secretion efficiency of different proteins fused to yeast signal sequences although size, net charge, and degree of glycosylation are all likely to affect secretion efficiency (26).

Gene Structure

As described earlier, secreted proteins have an amino-terminal extension responsible for translocation into the endoplasmic reticulum (ER) (27 ,28). This extension, termed the signal sequence, is usually composed of about twenty amino acids and is cleaved from the mature protein at some point after passage into the lumen of the ER (25). Signal sequences are usually comprised of three regions including a positively charged amino terminal region, a central hydrophobic core and a polar carboxy-terminal region containing a cleavage site(29 ,30). The importance of the hydrophobic core has been established in that at some point it plays a crucial role in the export of protein (31). The leader sequence acts at two stages of the export process: at entry into the secretion pathway and at the subsequent translocation across the ER membrane. How the structure of the signal sequence affects this process, in spite of the lack of amino acid consensus between signal sequences from different naturally secreted proteins, is the subject of a review by Randall (32).

The signal sequences most widely used in directing secretion in yeast come from the native *MF*α1 and *SUC*2 genes as summarized in Table 2. *S. cerevisiae* can exist in any one of three distinct cell types. Two haploid types designated **a** and α can fuse to form an **a**/α diploid cell. The process of **a** and α mating is initiated by the reciprocal exchange of pheromones, called a-factor and α-factor, respectively (33).

The α-factor has been fairly well characterized in regards to secretion. Its structure and secretion pathway have been elucidated (34 ,35). The *MF*α1 gene contains four tandem copies of mature α-factor within a putative polypeptide precursor of 165 amino acids termed pre-pro-α-factor. "Pre" refers to the amino terminal signal sequence composed of 22 mostly hydrophobic amino acids. This signal sequence resembles those found in a wide variety of secretory protein precursors and targets the polypeptide for processing through the secretion pathway (27). The signal sequence is followed by the "pro" section of the polypeptide which consists of approximately 60 amino acids and contains three possible glycosylation sites (34). Although the function of the "pro" section is unknown, it may be involved in targeting the α-factor precursor to other processing sites in the secretion pathway. Four tandem copies of α-factor follow the "pro" section, each preceded by spacer peptides which contain signals for processing by both endo and exoproteases (24). The *MF*α1 gene is regulated only by the sex of the cell either **a** or α and thus, in an α haploid, α-factor is constitutively produced (33).

Yeast strains containing any of the *SUC* genes (*SUC1-SUC7*) produce the enzyme invertase (36). SUC⁺ yeast strains produce an internal non-glycosylated form of invertase and a secreted glycosylated form of the enzyme (37). The internal form is produced constitutively, and the level of the external glycosylated enzyme is regulated by glucose repression. Increases up to 1000-fold have been shown in response to glucose derepression (38). The external form of the enzyme accounts for most of the activity when the cell is in the derepressed state, while the internal non-glycosylated form of the enzyme is only produced at low levels (38).

Carlson and colleagues (37) proposed a model for glucose regulation of the *SUC* gene. According to this model, a 1.9 kb regulated mRNA encodes the entire sequence for the precursor of secreted invertase. Translation of this mRNA begins with a methionine (met) codon at the beginning of the signal sequence. A 1.8 kb non-regulated mRNA begins within the signal sequence region and lacks the start-of-translation codon used for the 1.9 kb mRNA. As a result, translation begins at the next met codon and none of the secretion signal is translated.

Two promoter sites are present on the *SUC2* segment. One does not respond to glucose repression and results in the constitutive production of intracellular invertase. The other promoter is subject to glucose regulation and is responsible for the synthesis of secreted-glycosylated invertase (39). A scheme was developed whereby the *SUC2* signal sequence, which is composed of 20 mostly hydrophobic amino acids, was mutated in various places and returned to viable yeast cells (28). Results of this study confirm that the signal sequence is essential for proper secretion of the glycosylated form of invertase but in no way affects biological activity.

Invertase (β-D-fructofuranosidase fructohydrolase EC 3.2.1.26) catalyzes the hydrolysis of sucrose into its constituents, glucose and fructose. It has been shown that external, glycosylated invertase has a total weight of approximately 270,000 daltons, half of which is carbohydrate(40). The protein fraction of the enzyme is composed of two identical 60,000 dalton subunits, to each of which is added an average of nine neutral oligosaccharide chains (41). Internal invertase is similar in structure but is not glycosylated.

Previous Work

Many factors are proposed to have some influence on either the efficiency or the kinetics of the secretion process in yeast. Of primary concern are the elements of the gene cassette including the promoter, signal sequence, structural gene and terminator, the location of the cassette on either the host cell's chromosome or a plasmid, the number of copies of the cassette in the cell and the genetic characteristics of the host strain. Environmental factors which may influence the secretion process include media composition, the presence of substances which induce or repress the promoter, and various fermentation variables such as medium pH, dissolved oxygen content, and impeller speed. The mode of the fermentation, either batch, fed-batch or continuous is grouped with the other environmental variables since it can have a profound effect on growth and product formation and thus may influence the efficiency of the secretion process.

Signal Sequence Effect. Several authors have attempted to attain the secretion of heterologous proteins from yeast using fusions which contain the non-yeast signal sequence originally attached to the structural gene of interest. These constructions have had varied success in transporting their products out of the yeast cell. In most cases, the level of secreted product protein was low and the preprotein processing was inefficient, resulting in a mixture of products at several stages of internal processing. Table 1 shows some of the proteins for which the original secretion signal sequence was used.

A more successful method has been to include the signal sequence from a naturally secreted yeast protein in the gene cassette. This provides for entry into the secretion pathway, efficient processing, including cleavage of the signal sequence, and secretion of the heterologous protein. Based on this approach a number of fusions have been constructed (Table 2) which allow successful secretion into either the culture medium or the cell wall.

Since native yeast signal sequences seem to work most effectively the question arises as to which native yeast signal sequence works best. Emr (55) constructed fusions between the *SUC2* amino-terminal coding region including the promoter and signal sequence with the *E. coli lacZ* structural gene, which codes for the cytoplasmic enzyme β-galactosidase. These fusions produced hybrid invertase-β-galactosidase protein which exhibited β-galactosidase activity, and was regulated in the same manner as the wild type *SUC2* gene. Yet this hybrid protein was not secreted, but remained in the endoplasmic reticulum. Das (2) , on the other hand, constructed a fusion between the *MFα1* promoter/pre-pro-leader portion and the *E. coli lacZ* gene. Not only was this hybrid protein correctly processed but it was secreted into the yeast periplasmic space. Thus for the *E. coli* gene product β-galactosidase, the *MFα1* signal sequence seems to be the signal sequence of choice. This observation though, can not be generalized for other proteins. In an attempt to

TABLE 1. Foreign genes expressed in yeast employing their own signal sequence for secretion

Source	Protein	Reference
E. coli	β-lactamase	(42)
human	IFN-α/IFN-γ	(43)
wheat	α-amylase	(44)
Plant	thaumatin	(45)
mouse	immunoglobins	(46)
calf	prochymosin	(47)
human	α-antitrypsin	(48)
mouse	α-amylase	(49)
human	leukocyte-interferon-D	(43)

TABLE 2. Some typical gene fusions involving native yeast signal sequences that produce secreted protein

Promoter	Signal Sequence	Protein	Reference
*MFα*1	*MFα*1	human α-interferon	(26)
*MFα*1	*MFα*1	β-endorphin	(26)
*MFα*1	*MFα*1	calcitonin	(26)
*MFα*1	*MFα*1	invertase	(50)
SUC2	*SUC2*	invertase	(50)
SUC2	*SUC2*	calf prochymosin	(47)
GAL1	*SUC2*	calf prochymosin	(47)
TPI	*SUC2*	calf prochymosin	(47)
*MFα*1	*MFα*1	calf prochymosin	(47)
GAL1	*PHO5*	calf prochymosin	(47)
*MFα*1	*MFα*1	somatomedin-C	(51)
ACT	*MFα*1	somatomedin-C	(51)
CYC	*MFα*1	somatomedin-C	(51)
PHO5	*PHO5*	interferon	(52)
*ADH*1	*MFα*1	human lysozyme	(53)
*MFα*1	*MFα*1	human epidermal growth factor	(54)
*MFα*1	*MFα*1	*E. coli* β-galactosidase	(2)
SUC2	*SUC2*	*E. coli* β-galactosidase	(50)

express secreted calf prochymosin from yeast Smith et al. (47) constructed a series of gene fusions (some of which are shown in Table 2) incorporating various yeast promoters and signal sequences. The leader region from the *MFα*1 gene, the *PHO5* gene and the original calf prochymosin gene were all found to be less effective than the *SUC2* signal sequence in directing prochymosin secretion. Thus it appears that different signal sequences can be optimally effective for different proteins. This suggests that the physio-chemical properties of the product protein in question, play an important role in secretion efficiency.

Promoter Effect. Promoter strength is another variable which is able to affect secretion efficiency. In studies on secreted calf prochymosin (47) when the *SUC2* promoter was exchanged for the more powerful *TPI* promoter the expression level showed a five-fold increase, yet the amount of product secreted did not significantly increase. Moreover, secreted protein was found to contain both core and outer chain oligosaccharides while intracellular precursors contained only core glycosylations. This would suggest a limitation in the secretion pathway involving transit into, through or out of the Golgi complex. In a systematic study of the effect which the promoter has on secretion, Ernst (51) constructed three fusions each with a different promoter followed by the *MFα* signal sequence and the *SMC2* structural gene, which codes for the protein somatomedin-C. Each of these fusions was then placed on both a low copy number centromere vector (1 copy per cell) and a high copy number 2 μm-based vector. The promoters employed included those from the *CYC1, ACT* and *MFα1* genes which are relative to each other, weak, moderate and strong promoters. Experimental results showed that a strong promoter on a multicopy plasmid is not optimal for overall secretion. Instead, a constitutive promoter with a lower strength or a regulated promoter which is repressed in the early growth stages yielded best results for secretion. This suggests that high levels of protein expression may interfere with protein secretion as observed for bacteria (56).

Product Protein Effect. Zsebo and colleagues (26) have made fusions between the *MFα1* promoter/pre-pro coding region, and the various structural genes for the proteins β-endorphin, calcitonin, and consensus α-interferon (IFN-Con1) (26). Secretion efficiency, processing efficiency, and protein folding for these different sized proteins were studied for each cloned protein. β-endorphin and calcitonin are both small proteins, with lengths of 31 and 32 amino acids, respectively (57 ,58) , while interferon is a relatively longer protein with a length of 166 amino acids (59). The small proteins were efficiently secreted into the culture medium, while a substantial fraction of the larger protein (up to 90%) remained associated with the cell in the periplasmic space or cell wall. Manipulation of the processing site reduced processing efficiency but had no effect on secretion efficiency, suggesting that these two events are independent. It was also found that the secretion system as described above allows production of proteins with authentic amino termini and that the disulfide structure of the secreted interferon was identical to that of native human interferon. These attributes are significant in the preservation of biological activity of many proteins which might potentially be produced through fusions with the amino terminal end of naturally secreted yeast proteins.

Gene Location. Smith et al. (47) made the surprising observation that when the gene cassette was located on the chromosome, protein secretion was more efficient than when the same casette was located on a multicopy plasmid. When multiple copies of the casette were inserted on the chromosome, approximately the same absolute amount of protein was produced as found in strains harboring multicopy plasmids, yet four times more protein was secreted from the chromosomal strains. The reason for this is unclear, although it may have something to do with differences in either the rate or level of expression from chromosomal or plasmid-bourn gene copies.

Materials and Methods

The *Saccharomyces cerevisiae* model system employed in this study consisted of the host, strain SEY2102, transformed with the plasmids pSEY210 and pRB58 (37 ,50). This host is auxotrophic for uracil, contains a complete deletion of the chromosomal copy of the *SUC2* gene and contains no other invertase structural genes so that the host alone cannot produce invertase. The yeast 2μm based plasmids pSEY210 and pRB58 both contain the *ura*3 gene whose product protein is used as a selection marker. Plasmid pSEY210 contains the promoter and coding sequence for the "prepro" portion of the *MF*α1 gene, fused to the *SUC*2 structural gene. Plasmid pRB58 contains the yeast *SUC*2 promoter, signal sequence, and structural gene. Wild type yeast strain LBG H1022 was used as a control.

All strains were grown in selective medium: 6.7 g/L yeast nitrogen base without amino acids, 5.0 g/L casamino acids and either 20.0 or 2.0 g/L glucose. A fully automated 1.2 liter working volume fermentation system was used (Mouse, Queue Systems) which contained all the control devices and pumps required to maintain environmental parameters at their set points. A pH of 5.5, a temperature of 30°C and a dissolved oxygen content of 90% of maximum were maintained in all experiments. Seed cultures were grown in a shaker at 250 rpm, and the optical density of cell cultures was measured at 600 nm with a spectrophotometer.

In the transition experiments cells were first grown in a batch fermentor containing selective medium with an initial glucose concentration of 20 g/L. When glucose fell to approximately half of its initial value, cells were harvested by centrifugation, aseptically transferred to a second batch fermentor and grown in a fresh selective medium with an initial glucose concentration of 2 g/L.

An enzymatic colorimetric method was used to determine glucose concentration (Sigma, Glucose Procedure No. 510). A modified two step assay method (60) was used to measure invertase activity. (one unit was defined as the amount of enzyme which produced 1 μmol of glucose per minute at 30°C and pH 4.9.) Specific invertase activity was defined as units per ml of cell culture per optical density (A_{600}) of the culture.

In order to study protein localization in the yeast secretion process, a spheroplasting method previously described for wild type yeast cells(61) was modified for recombinant cells (62). Invertase activity in the cytoplasmic and periplasmic spaces was selectively fractionated by employing the enzyme zymolyase 20T (ICN Biomedicals), which was chosen for its ability to degrade yeast cell walls independently of cell age(63). Cells were grown to the desired cell density, harvested by centrifugation and washed with an equal volume of distilled water. The wet cell pellet was then resuspended in an osmotic buffer (1.0M mannitol/0.1M potassium phosphate/β-mercaptoethanol) pH 6.5 with a final zymolyase 20T concentration of 150 μg/ml. A reaction time of 60 minutes at 45°C was found adequate to degrade the cell walls leaving the plasma membrane intact. After 60 minutes spheroplasts were harvested by centrifugation (16,000xg, 30 seconds), and the supernatant was operationally defined as "periplasmic" proteins. The spheroplast pellet was resuspended in 0.5M phosphate buffer and mechanically lysed by intermittent vortexing with glass beads and cooling on ice. The lysate was centrifuged (16,000xg, 60 seconds) and the supernatant operationally defined as "cytoplasmic" proteins. The enzymatic activity of a yeast cytoplasmic marker protein glucose-6-phosphate dehydrogenase (G-6-PDH) was also followed using this procedure. The majority of G-6-PDH activity was found in the cytoplasmic fraction, suggesting that the fractions actually contain proteins from the spaces indicated.

Results

The objective of this investigation was to characterize several aspects of protein secretion in recombinant *S. cerevisiae*. Of particular interest was the effect of different promoter/signal-sequences and gene location on the localization of transported proteins and the kinetics of the secretion process. A time-trajectory of enzyme activity in the cytoplasmic, periplasmic and extracellular space, was determined for two recombinant strains and wild type yeast in an environmentally controlled fermentor. Both the *MFα1* and *SUC2* promoter/signal-sequences were fused to the *SUC2* structural gene which encodes the enzyme invertase. The efficiency of the transport process and the localization behavior of cloned invertase were compared between the two signal sequences (*MFα1* versus *SUC2*) and between the multicopy plasmid-harboring recombinant strain and the wild type. To analyze the transport kinetics of the *SUC2* gene in response to glucose concentration changes in the growth medium, transient experiments were carried out(64).

Promoter/Signal-Sequence   Effect. The   recombinant   yeast   strain (SEY2102/pRB58) containing multiple copies of the entire *SUC2* gene was grown in a selective medium with an initial glucose concentration of 20 g/L. Profiles of cytoplasmic, secreted and total specific invertase activity are shown in Figure 1. Secreted activity is the sum of that found in the periplasmic space and the culture broth. When the cells were in the exponential growth phase and the glucose concentration in the medium was high, both the cytoplasmic and secreted activity levels remained relatively constant with each accounting for approximately one half of the total activity.

After 12 hours of growth, the glucose concentration in the medium fell below 2 g/l, and the promoter for the *SUC2* gene was derepressed to produce more secreted invertase. Three hours after derepression, the specific enzyme activity in the periplasm reached a peak, six and a half times greater than its value in the exponential phase. By the end of the experiment, the periplasmic activity had fallen slightly and accounted for 59% of the total invertase produced (Table 3). In contrast, the specific activity transported into the culture broth rose linearly, but very slowly, for the entire time following derepression. At the end of the experiment it accounted for 18% of the total invertase produced. Invertase activity in the cytoplasm changed little, showing a peak at three hours after derepression followed by a slight decrease. It accounted for 23% of the total invertase produced at the end of the experiment.

Cells harboring plasmid pSEY210 which contain the *MFα1* promoter and prepro portion fused to the *SUC2* structural gene were grown under the same conditions as described above. The total invertase activity produced during the exponential phase was only slightly higher than that produced in the pRB58 strain (0.83 versus 0.72 units/ml·OD), yet the distribution (not shown) was quite different. The bulk of activity (48%) was found in the periplasmic space with the majority of the remainder (35%) located in the culture broth. As the experiment continued, the specific activity in the medium remained relatively constant while that in the periplasmic and cytoplasmic spaces began to increase. By the end of the experiment, the periplasmic

TABLE 3. Specific invertase localization at the end of batch experiments for three strains

	Specific invertase (unit/mL OD)		
	internal	secreted	total
wild type	0.06(14%)	0.38(86%)	0.44
SEY2102/pRB58	0.47(23%)	1.55(77%)	2.02
SEY2102/pSEY210	0.26(17%)	1.27(83%)	1.53

space activity level had reached 0.94 units/ml·OD and the total invertase activity represented a rise of 1.8 fold from the exponential phase of the experiment. Yet 17% of the total specific invertase activity was found in the cytoplasm over the entire course of the experiment. The reason for the higher total invertase level at the end of the experiment is not clear, though similar behavior was also observed for recombinant *E. coli* strains (Ryan, in press). Detailed studies are in progress to analyze the gene expression and protein localization pattern of the *MF*α1 promoter.

The *SUC2* promoter in the plasmid pRB58 was repressed by high glucose concentrations in the exponential phase of growth. As a result, the total amount of invertase produced was low and only about half was secreted out of the cytoplasm of the cell. After the glucose level dropped below 2 g/l, the total activity showed a 4-fold increase and the percentage secreted from the cytoplasm rose to 76%. In contrast, the *MF*α1 promoter in the plasmid pSEY210 yielded a constant percentage (83%) of total invertase secreted over the course of the experiment, regardless of the glucose concentration. Duplicate experiments performed with both the pRB58 and pSEY210 strains showed a consistent percentage of protein secreted, and hence consistent transport efficiency, for each of the two strains. The pRB58 strain produced 32% more total invertase activity than the pSEY210 strain. The small difference in the final percentage of secreted product between these two strains is probably insignificant, but the path to this result, which shows promoter dependent-behavior, is significantly different for each strain.

Gene Location Effect. The invertase localization profiles for wild type cells (not shown) resemble those for the pRB58 strain (Figure 1). When the glucose concentration in the medium was high, the invertase activity in the cytoplasm of the cells remained at a constant level (0.03 units/ml·OD), while the activity in the periplasmic space was too low to measure. When the glucose concentration fell below 2 g/l the invertase levels in the periplasmic and extracellular spaces reached respective peaks of 0.34 and 0.21 units/ml·OD, three hours after derepression. For each peak, this was followed by a slight decrease to a constant value of 0.24 and 0.14

units/ml·OD, respectively. Cytoplasmic activity remained relatively constant and only increased to 0.06 units/ml·OD at the end of the fermentation.

A comparison (Table 3) shows that the wild type secreted a larger percentage of its total specific invertase activity than the pRB58 strain. Yet the pRB58 strain produced over four times more total invertase activity. The larger amount of total invertase expressed by the pRB58 strain is accounted for in that this strain contains the *SUC2* gene on a multicopy plasmid, while the wild type has only one chromosomal copy of the gene. The culture broth of the pRB58 strain contained over twice as much invertase activity as that of the wild type, yet this amount of specific activity represented only about one fifth of the total produced in the pRB58 strain. For comparison, the culture broth of the wild type also contained one fifth of the total invertase activity it produced.

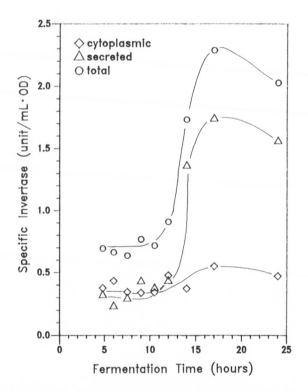

FIGURE 1. Profiles for cytoplasmic (◊), secreted (△), and total (○) invertase specific activity for strain SEY2102/pRB58 grown in a batch fermentation. Secreted invertase activity represents the sum of activities found in the periplasmic and extracellular spaces. Glucose concentration fell below 2 g/L, 12 hours into the fermentation.

Transition Experiments. Transition experiments were done to examine the localization of invertase activity and the transport kinetics of the *SUC2* gene after glucose in the medium decreased below the level required to repress the promoter. In these experiments, cells were initially grown in a selective medium containing 20 g/l glucose and then transferred, before derepression, to a fresh selective medium containing 2 g/l glucose. All other medium components were identical. This, in effect, distinguished a growth stage from a product expression stage.

The results obtained for the pRB58 transition experiment are shown in Figure 2. The invertase level in the periplasmic space reached a maximum three hours after transfer to the 2 g/L glucose medium. This level then gradually decreased with an accompanying, almost identical, increase in the extracellular space activity. The total secreted invertase activity, representing both the periplasmic and extracellular fractions, increased dramatically until three hours after derepression at which point it remained at a constant level for the rest of the experiment. This plateau coincides with the point at which the glucose in the fermentor was completely consumed and the concentration fell to zero. The cytoplasmic invertase activity (not shown) also showed a slight maximum at three hours after derepression and decreased to a constant level of 0.61 units/ml·OD by the end of the experiment (Table 4). The percentage of invertase secreted showed only a small change (from 74% to 80%) over the entire derepression stage of the experiment. Yet the fraction of activity actually found in the culture broth increased three-fold (from 7% to 24%) over this same period of time. It appears that this trend would have continued if the experiment had been carried out for a longer period of time. It seems that the large rise in activity accompanying derepression is rapidly secreted out of the cytoplasm into the periplasmic space, where it awaits a slower transport process to the culture broth. Although the shapes of the invertase activity profiles in both the transition and batch experiments are similar, the fresh medium in the second stage of the transition experiment allowed about 50% more total invertase activity to be produced.

A transition experiment was also performed with a wild type yeast strain nearly isogenic with the experimental strain. As observed in the pRB58 transition experiment, three hours after the start of derepression periplasmic invertase activity reached a maximum (0.49 units/ml·OD). The extracellular activity showed an accompanying linear increase (final value of 0.34 units/ml·OD). The cytoplasmic activity seemed to remain constant (0.06 units/ml·OD) for the entire derepression stage of the experiment. At the maximum activity level, 65% of the total invertase produced was secreted into the periplasmic space and 27% passed through the cell wall into the growth medium. At the end of the experiment, 52% of the total invertase produced had been secreted into the periplasmic space and 41% had passed through the cell wall into the growth medium. It is interesting that the pRB58 strain secreted over twice as much total invertase activity as the wild type into the culture broth. Yet this figure still represented only 24% of the total activity produced in the pRB58 strain which produced four times more total invertase activity than the wild type under the same conditions. Optical density did not decrease with derepression time in either the pRB58 or the wild type transition experiments. Preliminary results on cell size and protein content distributions measured by flow cytometry showed little

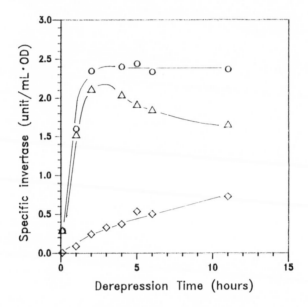

FIGURE 2. Profiles for periplasmic (△), extracellular (◇), and secreted (○) invertase specific activity for strain SEY2102/pRB58 grown in   the second stage of a transition experiment. Secreted invertase activity represents the sum of activities found in the periplasmic and extracellular spaces. Glucose in this stage of the experiment was completely consumed after 2 hours of derepression. Cell growth was nominal during this period.

TABLE 4. Specific invertase localization at the end of the transition experiments for wild type and recombinant yeast containing plasmid pRB58

| | Specific invertase (unit/mL OD) | | |
	internal	secreted	total
wild type	0.06(07%)	0.78(93%)	0.84
SEY2102/pRB58	0.61(20%)	2.37(80%)	2.98

change in these distributions during the derepression period, suggesting that the cells maintained their integrity.

Conclusion

The question of secretion efficiency can really be broken down into two issues especially for products like invertase which are reported to remain in the periplasmic space or cell wall. The first issue is the quantity of the recombinant product actually transported through the secretion pathway, past the cell membrane and secreted out of the cytoplasm. The second, and more important issue, is how much of this secreted product actually finds its way completely out of the cell, past the cell wall, into the culture broth. In an industrial setting the second issue is surely more important as it represents the product which is the easiest to attain in a purified form. With this in mind, invertase localization and transport kinetics were studied for wild type yeast and two recombinant *Saccharomyces cerevisiae* strains containing plasmids pRB58 and pSEY210. The plasmid pRB58 contains the *SUC2* promoter/signal-sequence fused to the structural gene for invertase. This promoter codes for external invertase which is repressed in the presence of high glucose concentrations. Plasmid pSEY210 contains the *MFα*1 promoter/signal-sequence fused to the structural gene for invertase and is known to produce secreted invertase independently of glucose concentration. The pRB58 strain showed a 3-fold jump in total invertase activity, a 6-fold increase in periplasmic invertase activity and a 2-fold jump in the amount secreted out of the cell into the medium when the glucose concentration dropped below 2 g/l. On the other hand, the pSEY210 strain showed a constant fraction (83%) of total invertase activity secreted out of the cytoplasm for the entire fermentation. The pSEY210 strain showed a slightly higher secretion efficiency than the pRB58 strain in that the former secreted 83% of its total invertase activity while the pRB58 strain secreted 77%; however, total invertase activity in the pRB58 strain was 33% higher than that in the pSEY210 strain.

The pRB58 recombinant strain which contains the *SUC2* gene on a multicopy plasmid produced four times more invertase activity than the wild type yeast strain which contains a single copy of the gene on a chromosome. At the maximum level of expression the wild type secreted 90% of the total invertase activity out of the cytoplasm, compared with 76% secretion in the pRB58 strain. These results coincide with the observations of Smith and colleagues (47) who found that a chromosomal copy of the *SUC2* gene is much more efficient in secretion and hence a higher percentage of its invertase is secreted. This may be due to two factors: first, the chromosomal site may for some reason be more efficient in targeting proteins for transport; or second, the overexpression of invertase in the multicopy-plasmid-harboring strain may interfere with the cellular transport machinery and thereby reduce secretion efficiency.

It is also interesting to compare *SUC2* gene expression under repressed conditions between the pRB58 strain, which contains multiple copies of the *SUC2* gene, and the wild type. when the pRB58 strain is compared to the wild type the glucose repression of secreted invertase seems "leaky". In the wild type fermentations little to no secreted invertase was produced when the glucose concentration in the medium was high. In contrast the pRB58 strain secreted nearly 50% of the total invertase produced even in high glucose concentrations. The abundance of gene copies in the plasmid-harboring strain may overwhelm the cells ability to repress the secreted form of invertase, leading to "leaky" expression in repressed conditions.

Transition experiments done with either the strain containing plasmid pRB58 or the wild type distinguished a growth stage from a product expression stage. Invertase activity in the periplasm reached a maximum approximately three hours after derepression, then slowly decreased with an accompanying gradual increase in the extracellular invertase. As stated earlier, the mechanism for protein transport through the cell wall is not as yet clear. It appears though, that during derepression expression and transport out of the cytoplasm occurred at a faster rate than passage across the cell wall, resulting in a rapid buildup of invertase activity in the periplasm accompanied by a slow rise in invertase activity in the extracellular space. Interestingly, optical density does not decrease with derepression time (data not shown). Preliminary results on cell size and protein content distributions measured by flow cytometry showed little change in these properties during the derepression period, suggesting that the cells maintain their integrity.

Acknowledgement

This research was supported by the National Science Foundation and the Petroleum Research Fund.

Literature Cited

1.   Schekman, R. and Novick, P., in *The Molecular Biology of the Yeast Saccharomyces: Metabolism and Expression*; Cold Spring Harbor Laboratory: Cold Spring Harbor, New York, 1982; pp. 361-393.

2.   Das, R. and Shultz, J., *Biotech. Prog.* 1987, **3**, 43-48.

3.   Arnold, Wilfred, *J. of Bacteriology* 1972, **112**, 1346-1352.

4.   Tkacz, J.S. and Lampen, J.O., *J. of Bacteriol.* 1973, **113**, 1073-1075.

5.   Levi, J.D., *Nature* 1956, **177**, 753.

6.   Scherer, B., *J. Bacteriol.* 1974, **119**, 386.

7.   Walter, P. and Blobel, G., *Proc. Natl. Acad. Sci.* 1980, **77**, 7112.

8.   Meyer, D.E., Krause, E., and Dobberstein, B., *Nature* 1982, **297**, 647.

9.   Hansen, W., Garcia, P.D., and Walter, P., *Cell* 1986, **45**, 397-406.

10. Waters, G. and Blobel, G., *J. of Cell Biol.* 1986, **102**, 1543-1550.

11. Deshaies, R.J., *Trends in Genetics* 1989, **5**, 87-93.

12. Novick, P., Feild, C., and Schekman, R., *Cell* 1980, **21**, 205-215.

13. Novick, P., Ferro, S., and Schekman, R., *Cell* 1981, **25**, 461-469.

14. Pfeffer, S. and Rothman, J., *Ann. Rev. Biochem.* 1987, **56**, 829-852.

15. Haselbech, A. and Schekman, R., *Proc. Natl. Acad. Sci.* 1986, **83**, 2017-2021.

16. Ferro-Novick, S., *TIBS*, 425-427.

17. Esmon, B., Novick, P., and Schekman, R., *Cell* 1981, **25**, 451-460.

18. Ballou, C.E., *The Molecular Biology of the Yeast Saccharomyces: Metabolism and Gene Expression*, 335. Cold Spring Harbor Laboratory,

19. Stevens, T., Esmon, B., and Schekman, R., *Cell* 1982, **30**, 439-448.

20. Franzusoff, A. and Schekman, R., *EMBO J.* 1989, **8**, 2695-2702.

21. Linnemas, W.A., Boer, P., and Elbers, P.F., *J. Bacteriol.* 1977, **131**, 638.

22. Field, C. and Schekman, R., *J. Cell. Biol.* 1980, **86**, 123.

23. Schekman, R., *Rec. Adv. Yeast Biol.* 1982, **1**, 1430.

24. Julius, D., Brake, A., Blair, L., Kunisawa, R., and Thorner, J., *Cell* 1984, **37**, 1075-1089.

25. Carter, B., Doel, D., Goodey, A.R., Piggott, J.R., and Watson, M.W., *Microbio. Sci.* 1985, **3**, 23-27.

26. Zsebo, K., Lu, H.S., Fischco, J., Goldstein, L., Davis, J., Duker, K., Suggs, S., Lai, P.H., and Bitter, G.A., *J. of Biolog. Chem.* 1986, **261**, 5858-5865.

27.   Blobel, G. and Doberstein, B., *J. of Cell Biol.* 1975, **67,** 835-851.

28.   Kaiser, C. and Bolstein, D., *Mol. and Cell. Biol.* 1986, **6,** 2382-2391.

29.   Perlman, D. and Halvarson, H. O., *J. Mol. Biol.* 1983, **167,** 391.

30.   VonHeijne, G, *J. Mol. Biol.* 1985, **184,** 99.

31.   Kaiser, C.A., Preuss, D., Grisati, P, and Botstein, D., *Science* 1987, **235,** 312.

32.   Randall, L.L. and Hardy, S.J.S., *Science* 1989, **243,** 1156-1159.

33.   Thorner, J., *The Molecular Biology of the Yeast Saccharomyces: Life Cycle and Inheritance,* 143. Cold Spring Harbor Laboratory,

34.   Kurjan, J. and Herskowitz, I., *Cell* 1982, **30,** 933-943.

35.   Julius, D., Schekman, R., and Thorner, J., *Cell* 1984, **36,** 309-318.

36.   Mortimer, R.K. and Hawthorn, C., *The Yeasts,* 385-460. Academic Press,

37.   Carlson, M. and Botstein, D., *Cell* 1982, **28,** 145-154.

38.   Gascon, S., Neumann, N.P., and Lampen, J.O., *J. of Biolog. Chem.* 1968, **243,** 1573-1577.

39.   Carlson, M., *Mol. Cell. Biol.* 1983, **3,** 439.

40.   Neumann, N.P. and Lampen, J.O., *Biochemistry* 1967, **6,** 468-475.

41.   Trimble, R. and Maley, F., *J. of Biololog. Chem.* 1977, **252,** 4409-4412.

42.   Roggenkamp, R., Kustermann-Kuhn, B., and Hollenberg, C.P., *Proc. Natl. Acad. Sci.* 1981, **78,** 4466.

43.   Hitzeman, R. A., Leung, D. W., Perry, L. J., Kohr, W. J., Levine, H. L., and Goeddel, D. V., *Science* 1983, **219,** 620-625.

44.   Rothstein, S.J., Lazarus, C. M., Smith, W. E., Baulcombe, D.C., and Gatenby, A.A., *Nature* 1984, **308,** 662.

45.   Edens, L., Bom, I., Lederboer, A.M., Maat, J., Toonen, M.Y., Visser, C., and Verrips, C.T., *Cell* 1984, **37,** 629.

46.   Wood, C.R., Boss, M.A., Kenton, J.M., Calvert, J.E., Roberts, N.A., and Emtage, J.S., *Nature* 1985, **314,** 446.

47.   Smith, R., Duncan, M., and Moir, D., *Science* 1985, **229,** 1219-1224.

48.   Cabezon, T., DeWild, M, Herion, P., Lorain, R., and Bollen, A., *Proc. Natl. Acad. Sci.* 1984, **81,** 6594.

49.   Thompson, K.K., *Carlsberg Res. Comm.* 1983, **48,** 545-555. _

50.   Emr, S., Schekman, R., Flessel, M., and Thorner, J., *Proc. Natl. Acad. Sci.* 1983, **80,** 7080 - 7084.

51. Ernst, J. F., *DNA* 1986, **5,** 483-491.

52. Hinnen, A., Meyhack, B., and Tsapis, R., *Proc. Alko. Yeast Sym.,
     Helsinki* 1983, **1,** 157.

53. Castanon, M. J., Spevak, W., Adolf, G. R., Chlebowicz-Sledziewska, E.,
     and Sledziewska, A., *Gene* 1988, **66,** 223-234.

54. Brake, A., *Proc. Natl. Acad. Sci.* 1984, **81,** 4642-4646.

55. Emr, S.D., Schauer, I., Aanson, W., Esmon, P., and Schekman, R., *Mol.
     Cell. Biol.* 1984, **4,** 2347-2355.

56. Emr, S.D., Schwartz, M., and Silhavy, T. J., *Proc. Natl. Acad. Sci.* 1978,
     75, 5802-5806.

57. Austin, B.M. and Smyth, A., *Biochem. Biophys. Res. Commun.* 1977,
     77, 86. 1977,

58. Jacobs, J.W., Goodman, R.H., and Chin, W.W., *Science* 1981, **213,**
     457-459.

59. Alton, K. and Stabinsky, B., , Elsevier Scientific Publishing Co.,
     Amsterdam, 1983.

60. Goldstein, A. and Lampen, J.O., *Meth. in Enzymol.* 1975, **42,** 504-511.

61. Wolska-Mitasko, B., *Analy. Biochem.* 1981, **116,** 241-247.

62. Marten, M.R. and Seo, J.H., *Biotech. Tech.* 1989, **3,** 325-328.

63. Kitamura, K., Kaneko, T., and Yamamoto, Y., *J. Gen. Appl. Microbiol.*
     1972, **18,** 57-71.

64. Marten, M.R. and Seo, J.H., *Biotech. Bioeng.* 1989, **34,** 1133-1139.

RECEIVED June 26, 1991

Chapter 8

Production of Antistasin Using the Baculovirus Expression System

D. Jain, K. Ramasubramanyan, S. Gould, C. Seamans, S. Wang, A. Lenny, and M. Silberklang

Merck Sharp and Dohme Research Laboratories, P.O. Box 2000, Rahway, NJ 07065

We have used *Spodoptera frugiperda* (Sf9) insect cells infected with recombinant baculovirus to produce antistasin, a potent anticoagulant and antimetastatic agent from leech salivary glands. Mature antistasin is a 119-a.a.,15-kDa secreted protein which has an unusually high cysteine content (20/119 residues) and acts by inhibiting coagulation Factor Xa. We have also genetically engineered a truncated (7.5 kDa) form of antistasin, "half antistasin" (H-ANS) for baculovirus-mediated expression. The efficiency of utilization of baculovirus in this system was improved by the use of a low (0.1) mutiplicity of infection. By monitoring increase in cell diameter during the course of infection and decrease in cell viability, both of which were found to correlate with product accumulation, harvest time could be optimized under various culture conditions. Secreted H-ANS activity peaked between 72 and 96 hours post-infection. Among the major process variables examined, dissolved oxygen at a low level of 10% and at a high level of 110% air saturation was found to result in a reduced growth rate as well as lower productivity relative to a level of 65%. Data are presented on the kinetics of oxygen uptake during cell growth and the viral infection cycle for the production of H-ANS.

The insect cell-baculovirus expression vector system is becoming increasingly popular for the production of recombinant proteins. Baculoviruses as a class have been well characterized because of

0097–6156/91/0477–0097$06.00/0

their potential applications as agricultural agents for insect pest control (1,2). This has provided an excellent basis both for the development of efficient expression vector systems using the abundantly expressed viral polyhedrin gene locus and for the production of viral and recombinant proteins in insect cell suspension culture (3-5). At Merck, we have used the baculovirus expression vector system to produce a number of recombinant proteins, one of which is antistasin, a potent anticoagulant and antimetastatic agent (6).

Antistasin is a protein found in the salivary glands of the Mexican leech *Haementaria officinalis* (7-9). It is a stoichiometric inhibitor of Factor Xa (10) in the blood coagulation pathway and has no homology to Hirudin (11-13), another leech (*Hirudo medicinalis*) derived anticoagulant which inhibits thrombin (14). Antistasin also inhibits experimentally induced tumor cell metastasis in a mouse model system (10,15). Antistasin is translated as a preprotein with a signal peptide of 17 amino acids (6). The mature protein has 119 amino acid residues, has no N-linked glycosylation sites and has a 17% cysteine content. There is a significant two-fold homology between the N- and C- terminal halves, with the active site residing in the N-terminal half (16).

Antistasin purified from natural leech populations contains multiple primary sequence variants (16); two of the variants have been cloned and expressed in the insect baculovirus expression vector system (6), as has a truncated gene expressing a portion of one variant ("half-antistasin", J.S. Tung et. al., manuscript in preparation). One of the variants was also expressed in yeast and in a mammalian cell line, GH3, a rat pituitary tumor line (17). The initial yields of antistasin variant 1 were highest in the insect-baculovirus system (about 1.7 fold better than in the yeast system and about 3.5 fold better than the mammalian system). The baculovirus system was therefore selected to study the production of antistasin variants as well as half-antistasin (H-ANS). In ongoing studies, both *in vivo* and *in vitro*, the baculovirus-produced homogeneous full antistasin protein has been found to be pharmacologically very similar to the leech-derived protein (C. Dunwiddie, P. Siegl and G.Vlasuk, personal communication).

A number of reports have been published on the large-scale cultivation of insect cells in batch and continuous suspension culture (18-20), in stirred fermenters (21) and in airlift fermenters (22-25). There have been reports that insect cells, like many animal cell lines, are shear-sensitive (agitation and/or gas sparging) (20,23,24,26,27), are affected by the presence of foam and air bubbles (22-24,28) and can be protected by the presence of Pluronic F68 (24,28-35) and by controlling the size of gas bubbles (25). It also has been reported that insect cells have a relatively high oxygen demand in comparison to most mammalian cells (25,36).

We initially developed a scalable insect baculovirus suspension culture process for the production of antistasin, which was subsequently found to work equally well for the genetically engineered H-ANS. In this paper we present studies on the production of H-ANS in batch suspension culture. Our investigation centered on the effect of dissolved oxygen (DO) level on the growth of the cells as well as the production of protein and on oxygen uptake rate (OUR) during product accumulation. We have used Pluronic F68 for its reported "protective" properties (29) and have verified experimentally that there is no deleterious effect of sparging on either cell growth or protein production under our conditions.

Materials and Methods

Cell line. A clonal isolate, Sf9 (M. Summers, unpublished), of the cell line IPLB-Sf21-AE (37) derived from pupal ovary tissue of the Fall Army Worm *Spodopetra frugiperda* was obtained from the ATCC (#CRL 1711) and used as cell source.

Virus. Recombinant *Autographa californica* nuclear polyhedrosis virus (AcMNPV) containing the gene for H-ANS was generated by standard procedures (3).

Cell culture medium. IPL-41 basal medium (from J.R. Scientific, Woodland, CA) with 2% heat-inactivated Fetal Bovine Serum (FBS, Gibco, Grand Island, NY), 3.3 g/l yeastolate (Difco, Detroit, MI) and 3.3 g/l lactalbumin hydrolysate (Difco, Detroit, MI) was used for both static and suspension cultures, except that 1.0 g/l Pluronic F68 (BASF Corporation, Parsippany, NJ) was added to suspension cultures.

Cell growth. A cryopreserved working seed stock of Sf9 cells was maintained in vials in liquid nitrogen in IPL-41 medium with 20% heat-inactivated FBS and 10% DMSO (Sigma, St, Louis, MO). Thawed cells first were grown in static culture in 75 cm^2 tissue culture flasks (Corning Glass, Corning, NY) in IPL-41/2% FBS medium at 28°C, and then transferred to suspension culture in spinner flasks in the same medium. For this investigation, all spinner flasks were the paddle-impeller microcarrier-type (Bellco Glass, Vineland, NJ) ranging in volume from 100 ml to 8 liters. 100 ml spinner flask cultures were seeded directly from tissue culture flask cultures and grown at 28 °C on slow-speed magnetic stir plates (Barnstead/ Thermolyne, Dubuque, Iowa) at 50-60 rpm. Suspension cultures were then propagated at a seeding density of 3-4x10^5 cells/ml. Maximal cell density was not allowed to exceed 2.0x10^6 cells/ml. Fresh suspension cultures were started every 4 weeks to eliminate the possibility of experimental variation due to the reported "aging effect" seen upon long-term passage of insect cells in suspension (38,39). For higher density cell culture, 250 ml and larger spinner-flasks were head-space gassed at about 0.1 volume/volume/min.; air was used as the starting gas and then a mixture of air and oxygen was used to keep the DO level between 50% and 100% air-saturation based on off-line measurements (see below).

Two fermenters were used for growth of insect cells, an 8-L "Biostat E" fermenter from B. Braun Biotech (Bethlehem, PA) utilizing a bubble-free silicone tube gassing system and an 18-l

Sulzer-MBR (Woodbury, NY) "Spinferm" using a 2 μm (pore size) stainless steel microsparger for direct sparging of oxygen. Virus stock production. Viral stocks were made by growing Sf9 cells in spinner flasks (3; Gibco, Grand Island, NY) and infecting them with recombinant baculovirus (AcMNPV) containing the gene for H-ANS. Cells were grown to a density of $1.0x10^6$ cells/ml using TNMFH/10% FBS medium, and virus was added at an MOI of 0.1 (e.g., 10 ml of virus stock having a viral titer of 10^7 plaque forming units (PFU)/ml added per liter of culture). Culture medium containing the virus was harvested between 72 and 96 hours post-infection (based on 50% loss of viability (trypan blue exclusion method; below) in the culture). Viral titer was obtained by end-point dilution (3).

Production of H-ANS. H-ANS was produced in both large (up to 5 liter) gassed spinner flasks and fermenters (described above). Cells were grown to a density of $1.5x10^6$ cells/ml in IPL-41/2% FBS medium and infected with virus. The cells were monitored for change in modal diameter and/or loss in viability (below). The culture was harvested within 24 hours of when the modal cell diameter peaked or when the cell viability dropped to 50%.

Calculation of oxygen uptake rate (OUR). OUR calculations in the Braun fermenter were made using the following procedure. The gas mixture supply (air, nitrogen and oxygen) to the fermenter was first turned off. In the case of the Braun fermenter, the silicone tubing was purged with nitrogen to eliminate any residual oxygen transfer during the measurement, and the exhaust was closed. For the Sulzer fermenter, the head space was purged with nitrogen and the exhaust was closed. The initial DO reading was noted. The drop in DO with time was observed and noted every 30 seconds for 5 minutes. The slope of the DO vs time curve was used to calculate the specific OUR.

Assay Methods. H-ANS activity was measured by a colorimetric Factor Xa inhibition assay using a commercial (Helena Laboratories, Beaumont, Texas) Factor Xa assay kit (G. Cuca, personal communication). The assay was performed in a 96-well micro-titer plate with a Bio-Rad (Richmond, CA) kinetic microplate reader. H-ANS activity is reported here as normalized relative units.

    Glucose, ammonia and lactate dehydrogenase were measured with an Ektachem Analyser Model DT60 equipped with a DTSC module (for enzyme analysis) from Kodak (Rochester, NY). Off-line DO measurements were made with a Blood Gas Analyser from Radiometer, Copenhagen. On-line measurements were made with a polarographic DO probe from Ingold (Wilmington, MA).

    Cell counts were made using a haemacytometer (AO Scientific Instruments, Buffalo, NY). Viability counts were made by a trypan-blue dye exclusion method. A 1:1 dilution of the sample was made with trypan-blue (0.4% in 0.85% saline, Gibco, Grand Island, NY), and cells that picked up the blue stain were scored as non-viable. Cell size distribution (modal cell diameter) was determined with a Coulter Counter (Hialeah, FL) Model ZBI

outfitted with a channeliser, and data were analysed on an IBM AT-
compatible computer. Particle sizes between 8 and 21 μm were
considered for cell size determination with the Coulter counter.

Results

The effect of sparging on Sf9 cell growth was studied by growing
cells in two different fermenters, one using bubble-free silicone
tube gassing (Braun) and the other using direct oxygen sparging
(Sulzer-MBR). A 250 ml spinner flask equipped with head-space
gassing was used as a control. Figure 1 compares growth of cells
in the three systems. Growth of cells was similiar, and there was
no apparent deleterious effect of sparging. Specific growth rate
was in the range of 0.024 to 0.026 hr^{-1} (cell doubling time of
about 27.5 hours).
      The level of DO in the culture medium had a dramatic
effect on the growth rate of cells in the first 90-100 hours of
growth, as shown in Figure 2. At a low DO level of 10% air
saturation, and at high DO level of 110% air saturation maximum
specific growth rate was slower (about 25%) than at a DO level of
65% air saturation. We had observed in initial experiments in
spinner flasks that the maximum specific growth rate is obtained
in the first 90-100 hours of growth. Growth rates in fermenters
were, therefore, not monitored beyond 100 hours. It is significant
that the viabilities seemed to be similiar (ca. 98%) at the three
DO conditions (data not shown), implying that the reduced growth
rate of the cells is probably not the result of an increase in
cell death but a direct effect of DO concentration on growth.
      In our early experiments with antistasin variant 1,
highest product accumulation was obtained when the multiplicity of
infection (MOI) was less than 1.0 (unpublished observation).
However, with the H-ANS baculovirus vector, peak product yield was
unaffected by MOI. A typical experiment for H-ANS production at
different MOI is shown in Figure 3; at an MOI of 0.1, 0.5, 2.0 or
5.0, product yields were similiar in 250 ml spinner flask
cultures. Thus, to conserve virus, all subsequent production
experiments used an MOI of 0.1.
      Since baculoviruses are lytic viruses, harvest time
selection can be critical for product yield. We had previously
examined monitoring of a number of parameters as indicators of
harvest time, in lieu of the relatively tedious off-line product
assay, and found a strong correlation (in spinner flasks) between
the increase in modal cell diameter (as measured by a Coulter
Counter) and antistasin production (40, S. Gould et. al.
manuscript in preparation). This correlation subsequently was
found to be consistent between experiments in spinner flasks and
in fermenters for H-ANS production. It was observed that,
typically, modal cell diameter peaked about 24 hours before
maximum antistasin or H-ANS production. Cell viability, though not
as accurate, is much more convenient to follow and also correlates
well with product accumulation. It was found in numerous
experiments that a 50% drop in cell viability coincided with the
peak in antistasin or H-ANS production. Figure 4 illustrates the
relation between the kinetics of H-ANS accumulation and increase

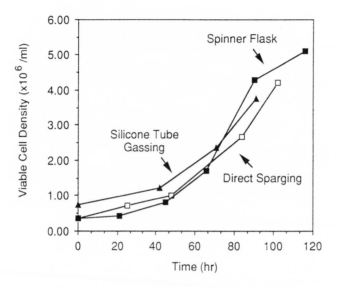

Figure 1. Cell growth with different gassing systems. The spinner flask was gassed via head-space only, while the fermenters utilized either silicone tube gassing (Braun) or direct microsparging (Sulzer-MBR). Cell viability remained above 95% throughout in all cases.

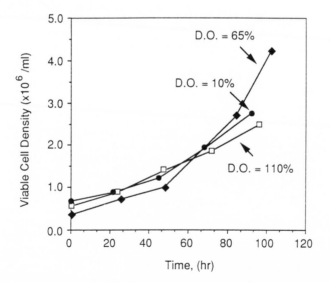

Figure 2. Effect of dissolved oxygen (DO) on cell growth. Experiments were run sequentially in the Sulzer-MBR 18-1 "Spinferm" fermenter with oxygen provided by microsparging of air/oxygen mixtures.

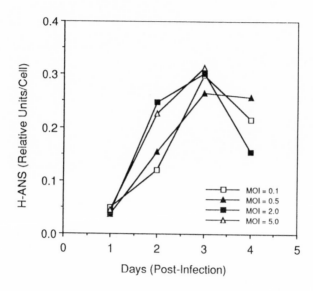

Figure 3.   Effect of multiplicity of infection (MOI) on H-ANS production. Viral stocks were prepared in TNMFH/10% FBS and added directly to cells grown in IPL41/2% FBS in parallel 250 ml spinner flasks.

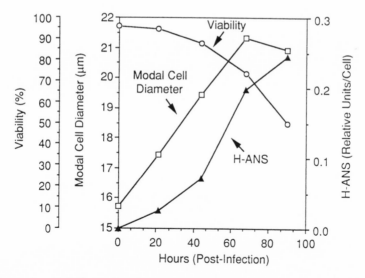

Figure 4.   H-ANS production as a function of cell diameter and cell viability during the infection cycle. The experiment was performed in the Sulzer-MBR 18-1 fermenter. Viability was measured by trypan blue exclusion and modal cell diameter in a Coulter Counter.

in modal cell diameter as well as cell viability in an 18-1
fermenter (Sulzer-MBR). It is clear that modal cell diameter
increases rapidly as H-ANS accumulates in the medium and peaks
about 24 hours before maximum H-ANS production (which, in turn,
coincides with a 50% loss in viability).

The effect of mode of oxygenation on production of H-ANS
is illustrated in Figure 5A. The same two fermenter systems, one
using bubble-free silicone tube gassing (Braun) and the other
using direct oxygen sparging (Sulzer-MBR) were used, with a 250 ml
head-space gassed spinner flask as control. H-ANS production was
found to be similiar and there was no apparent deleterious effect
of sparging. Figure 5B shows the cell viability in the three
systems during the infection cycle. There seems to be no
difference in the rate of cell death in the baculovirus-infected
cultures in the three systems.

The effect of DO level on production of H-ANS is shown in
Figure 6A. The experiment was done in the silicone tube-gassed
Braun fermenter to exclude any potential effect of sparging on
production. Here again, as for cell growth (Figure 2), it was
found that DO levels of 10% and 110% result in reduced
productivity of H-ANS as compared to a DO level of 65%. Figure 6B
illustrates the viability of the cells after baculovirus infection
at the three different DO levels. The viable count decreases much
more rapidly at a DO level of 110% than at 65% or at 10%. For
these experiments, cells were grown at 65% DO and then infected at
different DO concentrations.

The specific OUR during the production cycle of the
cultures at the three DO levels is shown in Figure 7. The OUR
value obtained at 65% DO is comparable to those reported in the
literature (25). It is interesting to note that the specific OUR
of the cells, which can be an indicator of their metabolic
activity, is greater at the higher DO levels. It is almost twice
as much at 65% DO than at 10% DO while at 110% DO it is more than
four times that at 10% DO.

Figure 8 presents a comparison of the viability drop in
the culture at 65% DO and the specific OUR of the cells. There is
a small increase (<10%) in the specific OUR upon infection of the
cells with virus early in all culture. We also see an apparent
rise in specific OUR (oxygen utilization per trypan blue negative
cell per second) late in the infection cycle (Figure 8).

Discussion

We have shown that various parameters of Sf9 cell growth and
infection and H-ANS production reproduce well from the spinner
flask to the 18-1 fermenter scale. The issue of sensitivity to
shear caused by the presence of bubbles has been effectively
eliminated by using silicone tubing or microsparging, and the
addition of Pluronic F68. Even during the infection cycle, when
the infected cells swell by almost 38% in diameter and are much
more fragile, there seems to be no significant increase in cell
death due to microsparging in the fermenter environment (Figure
5B).

The use of an MOI of less than 1.0 is interesting. An MOI

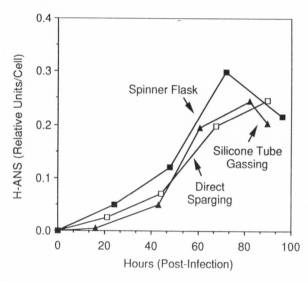

Figure 5A. H-ANS production under different gassing conditions.
The 250 ml spinner flask was gassed in the head-space only.
Silicone tube gassing was used in the Braun 8-1 fermenter and
direct microsparging in the Sulzer-MBR 18-1 fermenter.

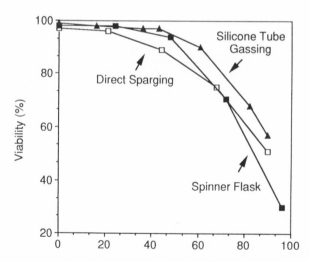

Figure 5B. Cell viability during H-ANS production under
different gassing conditions (as described in Figure 5A).

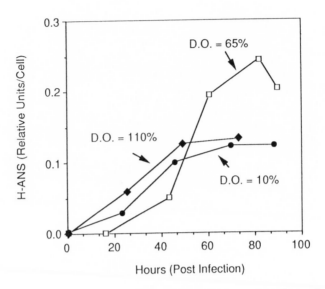

Figure 6A. Effect of dissolved oxygen (DO) on H-ANS production.
In this experiment, all 3 conditions were run separately in the
Braun 8-1 silicone tube gassed fermenter.

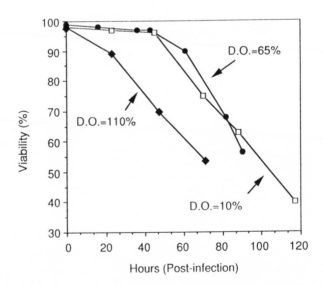

Figure 6B. Effect of dissolved oxygen (DO) on cell viability
during H-ANS production (as described in Figure 6A).

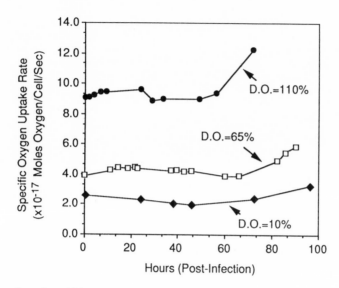

Figure 7. Specific oxygen uptake rates (OUR) at different dissolved oxygen levels. Specific OUR measurements were made in the Braun 8-1 fermenter during production of H-ANS under different dissolved oxygen conditions (as shown in Figure 6B).

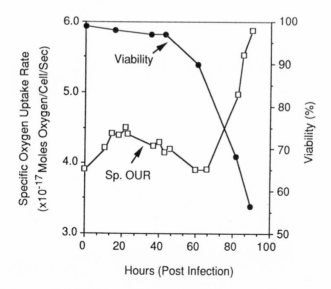

Figure 8. Viability and specific oxygen uptake rate during H-ANS production at 65% dissolved oxygen concentration (as shown in Figures 6 and 7).

of 0.1 should only infect, at best, 10% of the cells, while the uninfected cells should continue to grow and divide. We have, in fact, observed that with an MOI of 0.1, cell density can almost double after infection. Product assays were therefore calculated and are reported as relative units per peak cell number to compensate for this change. Data shown in Figure 3 support the presence of a delayed infection as H-ANS production for MOI's of 0.1 and 0.5 show a "lag" compared to MOI's of 2.0 and 5.0.

The effect of DO on cell growth and H-ANS production (Figure 2 and Figure 6A) is striking. Without more data, we can only hypothesize that cells (for the first 100 hours) grow slowly at 10% DO because they are oxygen starved while at 110% DO they are experiencing oxygen toxicity effects. In both situations cell death does not seem to increase. During the infection cycle, however, cell death is accelerated at 110% DO (Figure 6B). If we consider specific OUR primarily as a measure of metabolic activity we could hypothesize that cells are metabolically more active at higher DO levels and that the viral infection process is accelerated in proportion, leading to more rapid cell death. Alternatively, oxygen toxicity effects at DO levels above 100% air saturation could be more significant for infected than non-infected cells.

The rise in specific OUR late in the infection cycle (Figure 8) should be analysed with caution. The calculation of specific OUR was based on viable cell counts done using a trypan blue dye exclusion method. Our hypothesis is that some of the "blue cells" counted as dead cells were actually still respiring and contributing to the oxygen demand, resulting in the calculation of an artifically high specific OUR. This hypothesis merits experimental verification.

Conclusions

We have observed that various parameters of Sf9 cell growth, infection and H-ANS production scale up well from 250 ml spinner flasks to an 8-l silicone-tube gassed fermenter and an 18-l sparged (microsparger) fermenter. The progress of a baculovirus infection can be monitored effectively by following the increase in cell volume using a Coulter Counter and channelizer and also by following the decrease in cell viability. The production of H-ANS in our study was not dependent on the multiplicity of infection (MOI) between 0.1 and 5.0.

DO level in the culture medium has a strong influence on Sf9 cells. Very low (10%) and very high (110%) DO levels adversely affect cell growth (at least up to 100 hours in culture), infection and production. Good specific growth rate and productivity were obtained at 65% DO. We also observed that death of infected cells is accelerated at 110% DO. Specific OUR was found to increase with increase in DO. The specific OUR is almost twice as much at 65% DO as at 10% DO. At 110% DO the specific OUR is even higher. As has been reported by others (25), we observed that the specific OUR increased transiently upon baculovirus infection of the cells.

Acknowledgements

The authors acknowledge the generous help of L. O'Neill, K. Alves and J-S Tung with H-ANS gene expression, G. Cuca and C. Dunwiddie with antistasin assays, and G. E. Mark and R. W. Ellis for advice and support during the course of this work.

Literature Cited

1. Martignoni, M.E. In Chemical and Biological Controls in Forestry, Garner, W.Y. and Harvey, J., Ed.; American Chemical Society: 1984, p 55-67.
2. Huber, J. In The Biology of Baculoviruses, Granados, R.R and Federici, B.A., Eds.; CRC Press, 1986; p182-202.
3. Summers, M.D. and Smith, G.E. Texas Agricultural Experiment Section Bulletin, No. 1555, 1987.
4. Luckow, V.A., Summers, M.D. Bio/Technol. 1988, 6, 47-55.
5. Miller, L.K. Annu. Rev. Microbiol. 1988, 42, 177-199.
6. Han, J.H., Law, S.W., Keller, P.M., Kniskern, P.J., Silberklang, M., Tung, J-S., Gasic, T.B., Gasic, G.J., Friedman, P.A., Ellis, R.W. Gene. 1989, 75, 47-57.
7. Gasic, G.J., Viner E.D., Budzynski, A.Z., and Gasic, T.B.: Cancer Res. 1983, 43, 1633-1636.
8. Gasic, G.J., Iwakawa, A., Gasic, T.B., Viner, E.D., and Milas, L.: Cancer Res. 1984, 44, 5670-5676.
9. Murer, E.H., James, H.L., Budzynski, A.Z., Malinconico, S.M. and Gasic, G.J. Thromb. Haemostasis. 1984, 51, 24-26.
10. Dunwiddie, C., Thornberry, N.A., Bull, H.G., Sardana, M., Friedman, P.A., Jacobs, J.W., and Simpson, E. J. Biol. Chem. 1989, 264, 28, 16694-16699.
11. Chang, J.Y. FEBS Lett. 1983, 164, 307-313.
12. Harvey, R.P., Degryse, E., Stefani, L., Schamber, F., Cazenave, J.P., Courtney, M., Tolstoshev, P. and Lecocq, J.-P. Proc. Natl. Acad. Sci. USA. 1986, 83, 1084-1088.
13. Loison, G., Findeli, A., Bernard, S., Nguyen-Juilleret, M., Marquet, M., Riehl-Bellon, N., Carvallo, D., Guerra-Santos, L., Brown, S.W., Courtney, M., Roitsch, C. and Lemoine, Y. Bio/Technol. 1988, 6, 72-77.
14. Hemker, H.C.: In International Society on Thrombosis and Haemostasis, Verstraete, M., Vermylen, J., Lijnen, H.R., and Arnout, J., Eds. Leuven University Press, Leuven, 1987.
15. Tuszynski, G.P., Gasic, T.B., and Gasic, G.J. J. Biol. Chem. 1987, 262, 9718-9723.
16. Nutt, E., Gasic, T.B., Rodkey, J., Gasic, G.J., Jacobs, J.W., Friedman, P.A., and Simpson, E. J. Biol. Chem. 1988, 263, 10162-10167.
17. Silberklang, M., Kopchick, J., Munshi, S., Lenny, A., Livelli, T., and Ellis, R.W.: In Modern approaches to animal cell technology, Spier R.E. and Griffiths, J.B., Eds. European Society for Animal Cell Technology: The 8th Meeting. 1987, p199-214.
18. Kompier, R., Tramper, and J., Vlak, J.M. Biotechnology Lett. 1988, 10, 12, 849-854.

19. De Goojer C.D., van Lier, F.L.J., van der End, E.J., Vlak, J.M. and Tramper J. Appl. Microbiol. and Biotechnol. 1989, 30, 497-501.
20. Tramper, J. and Vlak, J.M. Ann. N.Y. Acad. Sci. 1986 469, 279-288.
21. Miltenburger, H.G. and David, P. Develop. Biol. Standard. 1980, 46, 183-186.
22. Tramper, J., Smit D., Straatman, and J., Vlak, J.M. Bioprocess Engineering.1988 3, 37-41.
23. Tramper, J., Joustra, D., Vlak, J.M.: In Plant and animal cell cultures: Process possibilities, Webb, C. and Mavituna, F., Eds. Chichester: Ellis Horwood, 1987; p125-136.
24. Tramper, J., Williams, J.B., Joustra, D. and Vlak, J.M. Enzyme and Microbiol. Technol. 1986, 8, 33-36.
25. Maiorella, B., Inlow, D., Shauger, A. and Harano, D. Bio/Technol. 1988, 6, 1406-1410.
26. Hink, W.F.: In Microbial and Viral pesticides, E. Kurstak, Ed. Marcel Dekker, Basel 1982; p 493-506.
27. Weiss, S.A., Kalter, S.S., Vaughn, J.L. and Dougherty, E. In Vitro 1980, 16, 222-223.
28. Handa, A., Emery, A.N. and Spier, R.E. Develop. Biol. Standard 1987, 66, 241-252.
29. Murhammer, D.W. and Goochee, C.F. Bio/Technol. 1988, 6, 1411-1418.
30. Swim, H.E. and Parker, R.F. Proc. Soc. Exp. Biol. Med. 1960, 103, 252-254.
31. Schmolka, I.R. J. American Oil Chemists Soc. 1977, 54, 110-116.
32. Reuveny, S., Velez, D., Macmillian, J.D., and Miller, L. J. Immunological Methods. 1986, 86, 53-59.
33. Radlett, P.J., Telling, R.C., Stone, C.J. and Whiteside, J.P. Applied Microbiology. 1971, 22, 534-537.
34. Kilburn, D.G. and Webb, F.C. Biotechnol. Bioeng. 1968, 10, 801-814.
35. Mizrahi, A. Develop. Biol. Standard. 1984, 55, 93-102.
36. Weiss, S.A., Orr, T., Smith, G.C., Kalter, S.S., Vaughan, J.L. and Dougherty, E.M. Biotechnol. Bioeng. 1982, 24, 1145-1154.
37. Vaughan, J.L., Goodwin, R.H. and Tompkins, G.J. In Vitro. 1977, 13, 213-217.
38. Hink, W.F. In Vitro. 1989, 25, 3, 24A.
39. Rechtoris, C. and McIntosh, A.H. In Vitro. 1976, 12, 10, p678-81.
40. Gould, S., Wang, S., Seamans, C., Lenny, A., Jain, D. and Silberklang, M. In Vitro. 1989, 25, 3, 47A.

RECEIVED January 16, 1991

INDEXES

Author Index

Affiliation Index

Subject Index

Production: Margaret J. Brown
Indexing: Janet S. Dodd
Acquisition: Robin Giroux and Barbara C. Tansill
Cover design: Neal Clodfelter

Printed and bound by Maple Press, York, PA